Progress in Mathematics
Vol. 41

Edited by
J. Coates and
S. Helgason

Birkhäuser
Boston · Basel · Stuttgart

Richard P. Stanley

Combinatorics and Commutative Algebra

1983

Birkhäuser
Boston • Basel • Stuttgart

Author:

Richard P. Stanley
Mathematics Department, 2-375
Massachusetts Institute of Technology
Cambridge, MA 02139

Library of Congress Cataloging in Publication Data

Stanley, Richard P., 1944-
Combinatorics and commutative algebra.

(Progress in mathematics ; v. 41)
Bibliography: p.
1. Commutative algebra. 2. Combinatorial analysis.
I. Title. II. Series: Progress in mathematics ; 41.
QA251.3.S72 1983 512'.24 83-17915
ISBN 0-8176-3112-7

CIP-Kurztitelaufnahme der Deutschen Bibliothek

Stanley, Richard P.:
Combinatorics and commutative algebra/
Richard P. Stanley. — Boston ; Basel ; Stuttgart :
Birkhäuser, 1983.
(Progress in mathematics ; 41)
ISBN 3-7643-3112-7 (Stuttgart, Basel)
ISBN 0-8176-3112-7 (Boston)

NE: GT

ISBN 0-8176-3112-7
ISBN 3-7643-3112-7
Printed in USA

A B C D E F G H I J

TABLE OF CONTENTS

PREFACE

These notes are based on a series of eight lectures given at the University of Stockholm during April and May, 1981. They were intended to give an overview of two topics from "combinatorial commutative algebra", viz., (1) solutions to linear equations in nonnegative integers (which is equivalent to the theory of invariants of a torus acting linearly on a polynomial ring), and (2) the face ring of a simplicial complex. In order to give a broad perspective many details and specialized topics have been regretfully omitted. In general, proofs have been provided only for those results which were obscure or inaccessible in the literature at the time of the lectures. The original lectures presupposed considerable background in commutative algebra, homological algebra, and algebraic topology. In order to broaden the accessibility of these notes, Chapter 0 has been prepared with the kind assistance of Karen Collins. This chapter provides most of the background information in algebra, combinatorics, and topology needed to read the subsequent chapters.

I wish to express my gratitude to the University of Stockholm, in particular to Jan-Erik Roos, for the kind invitation to visit in conjunction with the year devoted to algebraic geometry and commutative algebra at the Institut Mittag-Leffler. I am also grateful for the many insightful comments and suggestions made by persons attending the lectures, including Anders Björner, Ralf Fröberg, Christer Lech, and Jan-Erik Roos. Special appreciation goes to Anders Björner for the time-consuming and relatively thankless task of writing up these lecture notes. Finally I wish to thank Maura A. McNiff and Ruby Aguirre for their excellent preparation of the manuscript.

Richard Stanley
Cambridge, Massachusetts
May, 1983

NOTATION

$\mathbb{C}$	complex numbers
$\mathbb{N}$	nonnegative integers
$\mathbb{P}$	positive integers
$\mathbb{Q}$	rational numbers
$\mathbb{R}$	real numbers
$\mathbb{Z}$	integers
$\mathbb{R}^+$	nonnegative real numbers
$[n]$	for $n \in \mathbb{N}$, the set $\{1,2,\cdots,n\}$
N-matrix	a matrix whose entries belong to the set N
$N[x]$	polynomials in x whose coefficients belong to the set N
$N[[x]]$	formal power series in x whose coefficients belong to the set N
$\# S$	cardinality of the finite set S
$\lvert\circ\rvert$	cardinality or geometric realization, according to context
$T \subseteq S$	T is a subset of S
$T \subset S$	T is a subset of S and $T \neq S$
$\alpha > 0$	for a vector $\alpha = (\alpha_1,\cdots,\alpha_n) \in \mathbb{R}^n$, this means $\alpha_i > 0$ for all i
k^*	nonzero elements of the field k
kE	vector space over k with basis E
$\cong$	symbol for isomorphism
$\approx$	symbol for homeomorphism
$\oplus$, $\coprod$	direct sum (of vector spaces or modules)
im f	image $f(M)$ of the homomorphism $f: M \to N$
ker f	kernel of $f: M \to N$
vol $\mathcal{P}$	volume (= Lebesgue measure) of the set $\mathcal{P} \subseteq \mathbb{R}^n$
δ_{ij}	the Kronecker delta (= 1 if i=j, and = 0 if $i \neq j$)

CHAPTER 0

BACKGROUND

§1. Combinatorics.

The purpose of this introduction is to provide the reader with the relevant background from combinatorics, algebra, and topology for understanding of the text. In general the reader may prefer to begin with Chapter I and refer back to this chapter only when necessary. We assume the reader is familiar with standard (first-year graduate) material but has no specialized knowledge of combinatorics, commutative algebra, homological algebra, or algebraic topology.

We begin with a discussion of rational power series in one variable [St_5, §IV]. Let $F(x) = \sum_{n\geq 0} f(n)x^n \in \mathbb{C}[[x]]$ be a formal power series with complex coefficients. We say $F(x)$ is <u>rational</u> if there exist polynomials $P(x), Q(x) \in \mathbb{C}[x]$ for which $F(x) = P(x)/Q(x)$, i.e., $F(x)Q(x) = P(x)$ in the ring $\mathbb{C}[[x]]$. Without loss of generality we may assume $Q(0) = 1$. Define $\deg F(x) = \deg P(x) - \deg Q(x)$.

1.1 THEOREM. Let $\alpha_1, \alpha_2, \cdots, \alpha_d$ be a fixed sequence of complex numbers, $d \geq 1$ and $\alpha_d \neq 0$. The following conditions on a function $f\colon \mathbb{N} \to \mathbb{C}$ are equivalent:

(i)
$$\sum_{n\geq 0} f(n)x^n = \frac{P(x)}{Q(x)},$$

where $Q(x) = 1 + \alpha_1 x + \cdots + \alpha_d x^d$ and $P(x)$ is a polynomial in x of degree less than d.

(ii) For all $n \geq 0$,

$$f(n+d) + \alpha_1 f(n+d-1) + \cdots + \alpha_d f(n) = 0 . \qquad (1)$$

(iii) For all $n \geq 0$,

$$f(n) = \sum_{i=1}^{k} P_i(n)\gamma_i^{\,n} ,$$

where $1 + \alpha_1 x + \cdots + \alpha_d x^d = \prod_{i=1}^{k} (1-\gamma_i x)^{d_i}$, the γ_i's are distinct, and $P_i(n)$ is a polynomial in n of degree less than d_i.

Sketch of proof. Fix $Q(x) = 1 + \alpha_1 x + \cdots + \alpha_d x^d$. Define four complex vector spaces as follows:

$V_1 = \{f: \mathbb{N} \to \mathbb{C}$ such that (i) holds$\}$

$V_2 = \{f: \mathbb{N} \to \mathbb{C}$ such that (ii) holds$\}$

$V_3 = \{f: \mathbb{N} \to \mathbb{C}$ such that (iii) holds$\}$

$V_4 = \{f: \mathbb{N} \to \mathbb{C}$ such that $\sum_{n\geq 0} f(n)x^n$

$= \sum_{i=1}^{k} G_i(x)(1-\gamma_i x)^{-d_i}$, for some polynomials $G_i(x)$

of degree less than d_i, where γ_i and d_i have

the same meaning as in (iii)$\}$.

It is easily seen that $\dim V_j = d$ for $1 \leq j \leq 4$. One readily shows $V_1 \subseteq V_2$, $V_4 \subseteq V_1$, $V_4 \subseteq V_3$. Hence $V_1 = V_2 = V_3 (= V_4)$. □

We next consider rational functions $F(x) = P(x)/Q(x)$ with $\deg P \geq \deg Q$, i.e., $\deg F(x) \geq 0$.

1.2 PROPOSITION. Let $f: \mathbb{N} \to \mathbb{C}$ and suppose that $\sum_{n\geq 0} f(n)x^n = P(x)/Q(x)$ where $P, Q \in \mathbb{C}[x]$. Then there is a unique finite set $E_f \subset \mathbb{N}$ (the exceptional set of f) and a unique function $f_1: E_f \longrightarrow \mathbb{C}^* = \mathbb{C} - \{0\}$ such that the function $g: \mathbb{N} \to \mathbb{C}$ defined by

$$g(n) = \begin{cases} f(n), & \text{if } n \notin E_f \\ f(n) + f_1(n), & \text{if } n \in E_f , \end{cases}$$

satisfies $\sum_{n\geq 0} g(n)x^n = R(x)/Q(x)$, where $R \in \mathbb{C}[x]$ and $\deg R < \deg Q$. Moreover, assuming $E_f \neq \phi$ (i.e., $\deg P \geq \deg Q$), define $m(f) = \max\{i: i \in E_f\}$. Then:

(i) $m(f) = \deg P - \deg Q$,

(ii) $m(f)$ is the largest integer n for which (1) fails to hold,

(iii) Writing $Q(x) = \prod_1^k (1-\gamma_i x)^{d_i}$ as above, there are unique polynomials $P_1,\cdots, P_k$ for which (2) holds for n sufficiently large. Then $m(f)$ is the largest integer n for which (2) fails.

Sketch of proof. By the division algorithm for $\mathbb{C}[x]$, there are unique polynomials $L(x)$ and $R(x)$ with $\deg R < \deg Q$, such that

$$\frac{P(x)}{Q(x)} = L(x) + \frac{R(x)}{Q(x)} .$$

We then define E_f, $g(n)$, and $f_1(n)$ by

$$\sum_{n\geq 0} g(n)x^n = \frac{R(x)}{Q(x)}$$

$$E_f = \{i: \text{ the coefficient of } x^i \text{ in } L(x) \text{ is nonzero}\}$$

$$\sum_{n\in E_f} f_1(n)x^n = L(x) .$$

The proof follows easily. □

If $n \in E_f$, then we call $f(n)$ an exceptional value of f. Thus f has no exceptional values if and only if $\deg P(x)/Q(x) < 0$.

Two special cases of Theorem 1.1 will be of interest to us here.

1.3 COROLLARY. Let $f: \mathbb{N} \to \mathbb{C}$, and let $d \in \mathbb{N}$. The following three conditions are equivalent:

(i) $$\sum_{n\geq 0} f(n)x^n = \frac{P(x)}{(1-x)^{d+1}},$$

where $P(x) \in \mathbb{C}[x]$ and $\deg P \leq d$.

(ii) For all $n \geq 0$,

$$\sum_{i=0}^{d+1} (-1)^i \binom{d+1}{i} f(n+i) = 0 .$$

(In the calculus of finite differences this condition is written $\Delta^{d+1} f(n) = 0$.)

(iii) $f(n)$ is a polynomial function of n of degree at most d. (Moreover, $f(n)$ has degree exactly d if and only if $P(1) \neq 0$. In this case, the leading coefficient of $f(n)$ is $P(1)/d!$.) □

1.4 DEFINITION. A <u>quasi-polynomial</u> (known by many other names, such as "pseudo-polynomial" and "polynomial on residue classes") <u>of degree d</u> is a function $f:\mathbb{N}\to\mathbb{C}$ (or $f:\mathbb{Z}\to\mathbb{C}$) of the form

$$f(n) = c_d(n)n^d + c_{d-1}(n)n^{d-1} + \cdots + c_0(n),$$

where each $c_i(n)$ is a periodic function (with integer period), and where $c_d(n)$ is not identically zero. Equivalently, f is a quasi-polynomial if there exists an integer $N > 0$ (viz., a common period of $c_0, c_1, \ldots, c_d$) and polynomials $f_0, f_1, \ldots, f_{N-1}$ such that

$$f(n) = f_i(n) \quad \text{if} \quad n \equiv i \pmod N .$$

The integer N (which is not unique) will be called a <u>quasi-period</u> of f.

1.5 COROLLARY. The following conditions on a function $f:\mathbb{N}\to\mathbb{C}$ and integer $N > 0$ are equivalent:

(i) f is a quasi-polynomial of quasi-period N.

(ii) $$\sum_{n\geq 0} f(n)x^n = \frac{P(x)}{Q(x)},$$

where $P(x), Q(x) \in \mathbb{C}[x]$, every zero α of $Q(x)$ satisfies $\alpha^N = 1$ (provided $P(x)/Q(x)$ has been reduced to lowest terms), and $\deg P < \deg Q$.

(iii) For all $n \geq 0$,

$$f(n) = \sum_{i=1}^{k} P_i(n)\gamma_i^{\,n} ,$$

where each P_i is a polynomial function of n and each γ_i satisfies $\gamma_i^{\,N} = 1$. □

For instance (see Ch. I, Cor. 4.2), if

$$\sum_{n\geq 0} f(n)x^n = V(x)/(1-x)^a(1-x^2)^b ,$$

where $V(x) \in \mathbb{C}[x]$ and $\deg V < a + 2b$, then $f(n)$ is a quasi-polynomial of quasi-period two. Thus there are polynomials $g(n), h(n)$ for which

$$f(n) = g(n) + (-1)^n h(n) .$$

We next discuss the problem of solving linear homogeneous equations in nonnegative integers. Let Φ be an $r\times n$ $\mathbb{Z}$-matrix, and define $E_\Phi = \{\beta \in \mathbb{N}^n \mid \Phi\beta = 0\}$. Thus E_Φ is the set of solutions in nonnegative integers to the system $\Phi\beta = 0$ of homogeneous linear equations with integer coefficients. Clearly E_Φ forms a submonoid of $\mathbb{N}^n$, i.e., is closed under addition and contains 0. Chapter I is devoted primarily to the study of the monoid E_Φ and the related set $E_{\Phi,\alpha}$ of solutions to $\Phi\beta = \alpha$.

Hilbert deduced from his Basis Theorem (see Theorem 2.2 below) that E_Φ is a finitely-generated monoid. A simple direct proof was given by Gordan in 1900. If we define a formal power series

$$E_\Phi(x) = \sum_{\beta\in E_\Phi} x^\beta ,$$

where $x^\beta = x_1^{\beta_1}\cdots x_n^{\beta_n}$ and $\beta = (\beta_1,\cdots,\beta_n)$, then it was shown by Hilbert that $E_\Phi(x)$ is a rational function of x. In 1903 E.B. Elliott described an algorithm for computing $E_\Phi(x)$ and discussed some special cases. This algorithm was subsequently used by MacMahon [MM, Sections VIII-X] to investigate a wide variety of combinatorial problems. In particular, in [MM, Section VIII, Ch. VII] he computes the number $H_3(r)$ of 3×3 $\mathbb{N}$-matrices whose row and column sums are equal to r. (Such matrices are called <u>integer stochastic matrices</u> or <u>magic squares</u>.) Anand, Dumir, and Gupta [A-D-G] rediscovered MacMahon's result and made some conjectures (see Ch. I, Conj. 1.1(i)-(iii)) about the number $H_n(r)$ of $n\times n$ $\mathbb{N}$-matrices with line sum r. Stanley [St_1] proved the conjectures of Anand-Dumir-Gupta and generalized them to "magic labelings of graphs". This included a related conjecture of Carlitz on <u>symmetric</u> matrices with equal line sums. Stanley's proof was based upon the Elliott-MacMahon algorithm and the Hilbert syzygy theorem of commutative algebra (see Ch. I.11). In these notes we will prove these results using different tools from commutative algebra. An independent geometric proof, was also given by E. Ehrhart, Sur les carrés magiques, C.R. Acad. Sci. Paris <u>277</u>A (1973), 575-577.

Next we review some geometric notions. Two standard references are:

B. Grünbaum, Convex Polytopes, Wiley, London-New York-Sydney, 1967.

P. McMullen and G.C. Shephard, Convex Polytopes and the Upper Bound Conjecture, Cambridge University Press, 1971.

A <u>(convex) polytope</u> $\mathcal{P}$ is the convex hull of finitely many points in $\mathbb{R}^n$. Equivalently, $\mathcal{P}$ is a bounded intersection of finitely many half-spaces. A convex polytope is homeomorphic to a closed d-dimensional ball for some d, called the <u>dimension</u> of $\mathcal{P}$ and denoted $\dim \mathcal{P}$. A <u>supporting hyperplane</u> of $\mathcal{P}$ is an (affine) hyperplane H which intersects $\mathcal{P}$ and for which $\mathcal{P}$ is contained in one of the two closed half-spaces determined by H. The intersection $\mathcal{P} \cap H$ is a <u>face</u> of $\mathcal{P}$, and we also call the void set ϕ a face. A 0-dimensional face is a <u>vertex</u> of $\mathcal{P}$, and a $(d-1)$-dimensional face is a <u>facet</u>. Every face F of $\mathcal{P}$ is itself a convex polytope, and a face G of F is a face of $\mathcal{P}$. The <u>boundary</u> $\partial\mathcal{P}$ of $\mathcal{P}$ is the union of all faces of $\mathcal{P}$ except $\mathcal{P}$ itself. It concides with the notion of boundary in the usual topological sense.

A polytope $\mathcal{P}^*$ is <u>dual</u> to $\mathcal{P}$ if there is an inclusion-reversing one-to-one correspondence $F \to F^*$ between the faces F of $\mathcal{P}$ and F^* of $\mathcal{P}^*$. Every polytope $\mathcal{P}$ has a dual, which we won't bother to construct here.

A <u>polyhedral complex</u> is a collection Γ of polytopes in $\mathbb{R}^n$ satisfying:

(a) If $P \in \Gamma$ and F is a face of P, then $F \in \Gamma$, and

(b) If $P, P' \in \Gamma$, then $P \cap P'$ is a common face of P and P'.

A <u>(convex)</u> <u>cone</u> C is a subset of $\mathbb{R}^n$ which is closed under the operation of taking <u>nonnegative</u> linear combinations. A cone C is <u>polyhedral</u> if it is the intersection of finitely many closed half-spaces (which must all contain 0 in their boundary). The <u>dimension</u> $\dim C$ of a polyhedral cone C may be defined as the dimension of its linear span. We define a <u>face</u> of C as we did for polytopes, except that ϕ is never a face. If $\{0\}$ is a face, we call it the <u>vertex</u> of C. A 1-dimensional face is an <u>extreme ray</u>.

§2. <u>Commutative algebra and homological algebra</u>.

Our basic references are [A-M] and:

P.J. Hilton and U. Stammbach, A Course in Homological Algebra, Springer-Verlag, Berlin-Heidelberg-New York, 1971.

P.J. Hilton and Y.-C. Wu, A Course in Modern Algebra, Wiley, New York, 1974.

H. Matsumura, Commutative Algebra, second ed., Benjamin/Cummings, Reading, MA, 1980.

Hilton and Wu is recommended in particular as a relatively painless introduction to homological algebra.

All our rings A are commutative with identity 1. In particular, every subring of A contains 1.

2.1 DEFINITION. A ring A is said to be <u>noetherian</u> if it satisfies the following equivalent conditions.

(a) Every non-empty set of ideals in A has a maximal element.

(b) Every ascending chain $I_0 \subseteq I_1 \subseteq \cdots$ of ideals of A eventually stabilizes (i.e., $I_i = I_{i+1}$ for all large i).

(c) Every ideal of A is finitely-generated.

2.2 HILBERT BASIS THEOREM. If A is noetherian, then the polynomial ring $A[x]$ is noetherian.

2.3 EXAMPLE. Clearly any field k is noetherian. By repeated application of the theorem, $k[x_1, \cdots, x_n]$ is noetherian.

2.4 PROPOSITION. If A is noetherian and B is a homomorphic image of A, then B is noetherian. In other words, if A is noetherian, then so is A/I for any ideal I of A.

2.5 DEFINITION. (a) A ring A is an <u>(integral) domain</u> if it has no zero-divisors, i.e., if $xy = 0$ in A then $x = 0$ or $y = 0$.

(b) A is <u>reduced</u> if it has no nonzero nilpotent elements, i.e., if $x^n = 0$ for $x \in A$ and $n > 0$, then $x = 0$.

2.6 DEFINITION. The <u>radical</u> Rad (I) of an ideal I of A is the ideal of A defined by

$$\mathrm{Rad}(I) = \{x \in A \mid x^n \in I \text{ for some } n > 0\}.$$

2.7 DEFINITION. A <u>prime</u> ideal p of A satisfies: $p \neq A$ and if $xy \in P$, then $x \in p$ or $y \in p$.

REMARK. $\mathrm{Rad}(p) = p$ for all prime ideals p of A.

2.8 DEFINITION. Let S be a multiplicatively closed subset of A such that $1 \in S$. Then the <u>ring of fractions</u> of A with respect to S is

$$S^{-1}A = \left\{ \left[\frac{a}{s}\right] \,\middle|\, a \in A,\ s \in S \right\}$$

where $[\frac{a}{s}]$ is an equivalence class of fractions defined as follows:

$$\frac{a_1}{s_1} = \frac{a_2}{s_2} \quad \text{if} \quad (a_1 s_2 - a_2 s_1)t = 0 \quad \text{for some} \quad t \in S.$$

The usual rules of addition and multiplication of fractions make $S^{-1}A$ into a ring.

Let $f \in A$ and $S = \{1, f, f^2, \cdots\}$. Then $S^{-1}A$ is usually written A_f.

Let p be a prime ideal. Then $S = A - p$ is multiplicatively closed. $S^{-1}A$ is usually written A_p and is called "localization of A at p".

2.9 DEFINITION. An <u>A-module</u> M is an abelian group (with the group operation denoted by $+$) on which A acts linearly. That is, if a and b are in A and x,y are in M, then

$$a(x+y) = ax + ay$$

$$(a+b)x = ax + bx$$

$$(ab)x = a(bx)$$

$$1x = x$$

We write ax or xa interchangeably.

REMARK. Any ideal I of A is an A-module. In particular, A is an A-module.

REMARK. If A is a field, then an A-module is a vector space over A.

2.10 DEFINITION. A <u>free A-module</u> M is isomorphic to $\coprod_{j \in J} M_j$, where each M_j is isomorphic to A as an A-module. A <u>basis</u> for a free A-module M is a set B of elements of M such that each $u \in M$ can be uniquely written in the form

$$u = \sum_{e \in B} x_e e \,, \quad x_e \in A \,,$$

where all but finitely many $x_e = 0$. If $|B| = n < \infty$, then $M \cong A^n$. If A is noetherian then all bases have the same cardinality, called the <u>rank</u> of M.

2.11 DEFINITION. An A-module M is <u>finitely-generated</u> if there exist $u_1, \cdots, u_n$ in M such that

$$M = Au_1 + \cdots + Au_n \,,$$

i.e., every element u of M can be written (not necessarily uniquely) in the form $u = x_1u_1 + \cdots + x_nu_n$, $x_i \in A$.

2.12 PROPOSITION. M can be generated by n elements if and only if M is isomorphic to a quotient of the free A-module A^n.

Note that an A-module M has one generator (M is called cyclic) if and only if it is isomorphic to A/I for some ideal I of A. Although as an A-module A/I is generated by one element (viz.,1), as a ring it may require any number of generators.

2.13 DEFINITION. (a) The annihilator Ann M of the A-module M is the ideal of A given by

$$\text{Ann } M = \{x \in A \mid xM = 0\} .$$

(b) An element $x \in A$ is a non-zero-divisor on M if whenever $u \in M$ and $xu = 0$, then $u = 0$. In other words, the map $M \xrightarrow{x} M$ given by multiplication by x is injective

Now let A be a subring of a ring B. We say that B is a finite A-algebra if B is finitely-generated as an A-module, i.e., $B = x_1A + x_2A + \cdots + x_nA$ for some finite set of elements $x_1, \cdots, x_n$ of B (see Def. 2.11). We also say that B is a finitely-generated A-algebra (or of finite type over A) if there are finitely many elements $x_1, \cdots, x_n$ of B such that every element of B can be written as a polynomial in $x_1, \cdots, x_n$ with coefficients in A. We then write $B = A[x_1, \cdots, x_n]$. Finally we say that B is integral over A if every element of B is the root of a monic polynomial whose coefficients belong to A. We then have:

$$\text{finite type} + \text{integral} = \text{finite.}$$

For instance, if B is a finitely-generated algebra over a field $k \subseteq B$ and A is a subalgebra of B, then B is automatically of finite type over A (since B is of finite type over $k \subseteq A$). Thus in this situation, integral = finite.

2.14 NOETHER NORMALIZATION LEMMA. Let k be a field and $A \neq 0$ be a finitely-generated k-algebra. Then there exist elements $y_1, \cdots, y_r \in A$ such that $y_1, \cdots, y_r$ are algebraically independent over k and A is integral over $k[y_1, \cdots, y_r]$.

A version of this result for graded algebras appears in Ch. 1, Lemma 5.2.

2.15 DEFINITION. Let M_1, M_2, M_3 be A-modules, and

$$f_1 : M_1 \to M_2$$
$$f_2 : M_2 \to M_3$$

be A-module homomorphisms. Then

$$0 \longrightarrow M_1 \xrightarrow{f_1} M_2 \xrightarrow{f_2} M_3 \longrightarrow 0$$

is a <u>short exact sequence</u> if f_1 is injective, f_2 is surjective and $\operatorname{im} f_1 = \ker f_2$. (Hence $M_3 \cong M_2/M_1$.)

More generally, define a sequence of A-module homomorphisms

$$\cdots \longrightarrow M_1 \xrightarrow{f_1} M_2 \xrightarrow{f_2} M_3 \xrightarrow{f_3} M_4 \longrightarrow \cdots$$

to be a <u>(long) exact sequence</u> if $\operatorname{im} f_i = \ker f_{i+1}$ for all i.

Three useful properties of short exact sequences $0 \to M_1 \to M_2 \to M_3 \to 0$ are the following:

(a) M_2 satisfies ACC if and only if M_1 and M_3 do. Here ACC denotes the <u>ascending chain condition</u>: every ascending chain $N_0 \subseteq N_1 \subseteq \cdots$ of A-submodules of an A-module N eventually stabilizes. Note that A itself satisfies ACC (as an A-module) if and only if A is noetherian.

(b) M_2 satisfies DCC if and only if M_1 and M_3 do. Here DCC denotes the <u>descending chain condition</u>: every descending chain $N_0 \supseteq N_1 \supseteq \cdots$ of A-submodules of N eventually stabilizes. A module or ring satisfying DCC is called <u>artinian</u>.

Note: If a ring A (commutative, with identity) satisfies DCC, then it satisfies ACC, but not conversely. However, an A-module may satisfy DCC but not ACC. For instance, the set $M = k[x^{-1}]$ of polynomials in x^{-1} over a field k has an obvious structure as a module over $k[x]$ (set $x^a = 0$ in M for $a>0$). Then M satisfies DCC but not ACC.

(c) Let M_1, M_2, M_3 be finitely-generated. If $0 \to M_1 \to M_2 \to M_3 \to 0$ is exact, then

$$\text{rank } M_1 - \text{rank } M_2 + \text{rank } M_3 = 0 ,$$

where rank M is the largest n for which M contains a submodule isomorphic to A^n.

2.16 DEFINITION. Let I be an ideal of the ring A satisfying $\cap I^n = (0)$. We have the quotient maps $p_{n+1}: A/I^{n+1} \longrightarrow A/I^n$, $n \geq 1$. A sequence $(x_1, x_2, \cdots)$ with $x_n \in A/I^n$ is coherent if $p_{n+1}x_{n+1} = x_n$ for all $n > 1$. There is an obvious ring structure that can be put on the set of coherent sequences. The resulting ring is the I-adic completion of A, denoted $\hat{A}$. We may identify A with the subring of $\hat{A}$ consisting of all eventually constant coherent sequences.

Similarly, if M is an A-module, then we have quotient maps $q_{n+1}: M/I^{n+1}M \longrightarrow M/I^nM$, $n \geq 1$. We define the I-adic completion $\hat{M}$ of M exactly analogously to $\hat{A}$. Then $\hat{M}$ has the natural structure of an $\hat{A}$-module.

For example, if $A = k[x_1, \cdots, x_m]$, then $\hat{A} \cong k[[x_1, \cdots, x_m]]$ More generally:

2.17 PROPOSITION. Let A be noetherian and $I = x_1A + \cdots + x_mA$. Then

$$\hat{A} \cong A[[X_1, \cdots, X_m]]/(X_1 - x_1, \cdots, X_m - x_m) .$$

2.18 DEFINITION. A category C consists of:

(a) A class of objects $A, B, C, \cdots$

(b) To each pair of objects A, B of C, a set $C(A,B)$ of morphisms from A to B. If $f \in C(A,B)$, then write $f: A \to B$.

(c) To each triple A, B, C of C, a law of composition

$$C(A,B) \times C(B,C) \to C(A,C) \ ,$$

where we write $gf = h$ if $(f,g) \mapsto h$, subject to the axioms:

(a_1) The sets $C(A_1,B_1)$ and $C(A_2,B_2)$ are disjoint unless $A_1 = A_2$ and $B_1 = B_2$,

(a_2) Given $f:A \to B$, $g:B \to C$, $h:C \to D$, then $h(gf) = (hg)f$.

(a_3) To each object A there is a morphism $1_A:A \to A$ such that for any $f:A \to B$ and $g:C \to A$, we have $f1_A = f$ and $1_Ag = g$.

The main category of concern to us will be C = A-Mod, whose objects are A-modules and whose morphisms are A-module homomorphisms.

2.19 DEFINITION. A (covariant) functor $F:C \to \mathcal{D}$ between categories C and $\mathcal{D}$ is a rule associating with each object X in C an object FX in $\mathcal{D}$, and with each morphism $f:X \to Y$ in C a morphism $Ff:FX \to FY$ in $\mathcal{D}$ such that

$$F(fg) = (Ff)(Fg)$$

$$F1_X = 1_{FX} \ .$$

For a contravariant functor, we have $Ff:FY \to FX$ and $F(fg) = (Fg)(Ff)$.

2.20 DEFINITION. Let M, N be A-modules. Let C be the free A-module $A^{M \times N}$ (i.e. the free module with one generator for each element of $M \times N$). Let D be the submodule of C generated by all elements of the following types:

$$(x+x',y) - (x,y) - (x',y)$$

$$(x,y+y') - (x,y) - (x,y')$$

$$(ax,y) - a(x,y)$$

$$(x,ay) - a(x,y)$$

Let $T = {}^{C}/_{D}$. For each basis element (x,y) of C, let $x \otimes y$ be its image in T. Then T is the tensor product of M and N, denoted $M \otimes N$ or $M \otimes_A N$.

By construction, we have $g: M \times N \to M \otimes N$ such that $g(x,y) := x \otimes y$ is an A-bilinear map. The tensor product satisfies the following universal property: Let P be an A-module. Then if $f: M \times N \to P$ is an A-bilinear map, there exists $h: M \otimes N \to P$, an A-bilinear map, that makes the following diagram commute:

$$\begin{array}{ccc} M \times N & \xrightarrow{f} & P \\ {\scriptstyle g}\downarrow & \nearrow {\scriptstyle h} & \\ M \otimes N & & \end{array}$$

If we fix A and M, then the map $F:$ <u>A-Mod</u> $\to$ <u>A-Mod</u> given by $F(N) = M \otimes_A N$ (or $N \otimes_A M$) is a covariant functor. In other words, given $f: X \to Y$ there is a (canonical) way of defining $M \otimes f: M \otimes X \to M \otimes Y$ satisfying the definition of functor, viz.,

$$(M \otimes f)(u \otimes x) = u \otimes f(x) .$$

2.21 EXAMPLE. (a) Let $M = N = A = k[x]$, where k is a field (or any ring). Then $M \otimes_k N \cong k[x,y]$, while $M \otimes_A N \cong A$.

(b) For any A-module M and ring A, we have that $M \otimes_A A$ is isomorphic to M via $m \otimes a \mapsto ma$.

(c) If A is an algebra over a field k and K is an extension field of k (so K is a k-vector space), then by "extending the scalars of A to K" we mean forming the K-algebra $A \otimes_k K$.

(d) If S is a multiplicatively closed subset of A containing 1, then we define the <u>module of fractions</u> $S^{-1}M$ with respect to S of the A-module M by

$$S^{-1}M = M \otimes_A S^{-1}A .$$

We may regard $S^{-1}M$ either as an A-module or $S^{-1}A$-module.

(e) If I is an ideal of A and $\hat{A}$ the I-adic completion of A, then the I-adic completion $\hat{M}$ of M is given by

$$\hat{M} \cong M \otimes_A \hat{A} .$$

2.22 DEFINITION. An A-module P is projective if for every surjective homomorphism $f:M \to N$ and homomorphism $g:P \to N$, there exists a homomorphism $h:P \to M$ making the following diagram commute:

$$\begin{array}{ccccc} M & \xrightarrow{f} & N & \longrightarrow & 0 \\ & {\scriptstyle h}\nwarrow & \uparrow{\scriptstyle g} & & \\ & & P & & \end{array}$$

Equivalently P is projective if every short exact sequence $0 \to L \xrightarrow{\alpha} L' \xrightarrow{\beta} P \to 0$ splits, i.e., $L' \cong L \oplus P$ such that $\alpha(u) = (u,0)$, $\beta(u,v) = v$. This turns out to be the same as saying that P is a direct summand of a free module.. Hence:

2.23 PROPOSITION. Free modules are projective.

2.24 THEOREM. Let $A = k[x_1, \cdots, x_n]$, where k is a field. Then projective A-modules are free.

Note. This theorem, proved independently by Quillen and Suslin, solves a famous problem of Serre. We will only be concerned with graded A-modules (see Ch. I.2). In this case, the theorem that graded projective modules over $k[x_1, \cdots, x_n]$ are free is much easier to prove and was known from the beginnings of homological algebra.

2.25 DEFINITION. (a) A chain complex C over a ring A is a sequence $C = \{C_q, \partial_q\}$ of A-modules C_q and homomorphisms $\partial_q : C_q \to C_{q-1}$ such that $\partial_q \partial_{q+1} = 0$. This is denoted

$$C: \cdots \longrightarrow C_{q+1} \xrightarrow{\partial_{q+1}} C_q \xrightarrow{\partial_q} C_{q-1} \longrightarrow \cdots .$$

Since $\partial_q \partial_{q+1} = 0$, we have $\operatorname{im} \partial_{q+1} \subseteq \ker \partial_q$. Define the q-th homology group of C by

$$H_q(C) = \ker \partial_q / \operatorname{im} \partial_{q+1} .$$

(b) A cochain complex C over A is a sequence $C = \{C_q, \delta_q\}$ of A-modules C_q and homomorphisms $\delta_q : C_q \to C_{q+1}$ such that

$\delta_q \delta_{q+1} = 0$. This is denoted

$$C: \cdots \longrightarrow C_{q-1} \xrightarrow{\delta_{q-1}} C_q \xrightarrow{\delta_q} C_{q+1} \longrightarrow \cdots \quad .$$

Define the q-th cohomology group of C by

$$H^q(C) = \ker \delta_q / \mathrm{im}\ \delta_{q-1} \quad .$$

Note that the difference between chain complexes and cochain complexes, and between homology and cohomology, is purely formal. Every cochain complex can be converted to a chain complex by reindexing.

2.26 DEFINITION. A projective (resp., free) resolution of an A-module M is an exact sequence

$$\mathcal{P}: \cdots \to P_1 \to P_0 \to M \to 0$$

of projective (resp., free) A-modules P_i. (It is easily seen that projective resolutions of M always exist.)

2.27 DEFINITION. Let M be an A-module, and let $\mathcal{P}$ be a projective resolution of M as above. If N is another A-module, then we have a chain complex

$$\mathcal{P} \otimes N: \cdots \to P_{n+1} \otimes N \xrightarrow{\partial_{n+1} \otimes 1} P_n \otimes N \xrightarrow{\partial_n \otimes 1} \cdots \to P_0 \otimes N \xrightarrow{\partial_0 \otimes 1} M \otimes N \to 0 \; .$$

Define

$$\mathrm{Tor}_n^A(M,N) = H_n(\mathcal{P} \otimes N) = \ker(\partial_n \otimes 1)/\mathrm{im}(\partial_{n+1} \otimes 1) \; .$$

The A-module $\mathrm{Tor}^n_A(M,N)$ does not depend, up to isomorphism, on the choice of projective resolution of M. Moreover, both $\mathrm{Tor}^n_A(M,-)$ and $\mathrm{Tor}^n_A(-,N)$ are covariant functors. Note that $\mathrm{Tor}_0^A(M,N) \cong M \otimes_A N$. A basic property of Tor is the isomorphism $\mathrm{Tor}_n^A(M,N) \cong \mathrm{Tor}_n^A(N,M)$.

2.28 DEFINITION. If M and N are A-modules, then $\mathrm{Hom}_A(M,N)$ denotes the set of all A-module homomorphisms $f: M \to N$. The set $\mathrm{Hom}_A(M,N)$ has the structure of an A-module via

$$(xf)(u) = x(f(u)),$$

for $x \in A$, $f \in \mathrm{Hom}_A(M,N)$, $u \in M$. If M and N are free A-modules of ranks m and n, then one can identify in an obvious way (after choosing bases for M and N) elements of $\mathrm{Hom}_A(M,N)$ with $m \times n$ matrices over A. We also set $M^* = \mathrm{Hom}_A(M,A)$. If M is free with basis $u_1,\cdots,u_m$, then M^* is free with dual basis $u_1^*,\cdots,u_m^*$ defined by $u_i^*(u_j) = \delta_{ij}$.

Next we note that $\mathrm{Hom}_A(-,N)$ is a contravariant functor. Namely, if $f: X \to Y$ is a homomorphism of A-modules, then define

$$f^*: \mathrm{Hom}_A(Y,N) \to \mathrm{Hom}_A(X,N)$$

as follows: given $g: Y \to N$ and $u \in X$, let $(f^*g)(u) = gf(u)$.

2.29 DEFINITION. With $\mathcal{P}$, M, N as in Def. 2.27, we have a cochain complex:

$$\mathrm{Hom}_A(\mathcal{P},N): \cdots \leftarrow \mathrm{Hom}(P_{n+1},N) \xleftarrow{\partial_{n+1}^*} \mathrm{Hom}(P_n,N) \xleftarrow{\partial_n^*}$$

$$\cdots \leftarrow \mathrm{Hom}(P_0,N) \xleftarrow{\partial_0^*} \mathrm{Hom}(M,N) \leftarrow 0 .$$

Define

$$\mathrm{Ext}_A^n(M,N) = H^n(\mathrm{Hom}_A(\mathcal{P},N)) = \ker \partial_{n+1}^* / \mathrm{im}\ \partial_n^* .$$

Again, the A-module $\mathrm{Ext}_A^n(M,N)$ does not depend, up to isomorphism, on the choice of $\mathcal{P}$. Moreover, $\mathrm{Ext}_A^n(-,N)$ is a contravariant functor. Note that $\mathrm{Ext}_A^0(M,N) = \mathrm{Hom}_A(M,N)$. (Also, $\mathrm{Ext}_A^n(M,-)$ is a covariant functor, but we will not need this fact.)

2.30 DEFINITION. An A-module I is *injective* if for every homomorphism $f:M \to I$ and *injective* homomorphism $g:M \to N$, there exists a homomorphism $h:N \to I$ making the following diagram commute:

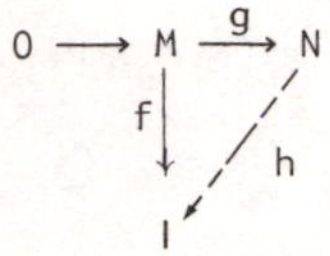

In other words, homomorphisms into I can be extended (from M to the larger module N which may be regarded as containing M). Equivalently, I is injective if it is a direct summand of every module which contains it.

2.31 DEFINITION and THEOREM. Given an A-module M, there is a unique (up to isomorphism) injective A-module $E_A(M)$ containing M, with the property that every injective A-module containing M also contains $E_A(M)$. $E_A(M)$ is called the *injective hull* or *injective envelope* of M .

2.32 EXAMPLE. Let $A = k[x_1,\cdots, x_n]$, where k is a field. Regard k as an A-module *via* the isomorphism $k \cong A/(x_1A +\cdots+ x_nA)$. Then $E_A(k) \cong k[x_1^{-1},\cdots, x_n^{-1}]$. Note that $E_A(k)$ is not finitely-generated.

2.33 DEFINITION. An *injective resolution* of an A-module M is an exact sequence

$$\mathcal{I}: 0 \to M \to I_0 \to I_1 \to \cdots ,$$

where each I_i is an injective A-module. (It is easily seen that injective resolutions of M always exist.)

2.34 DEFINITION. Let $\mathcal{C}$ = *A-Mod*. A functor $F: \mathcal{C} \to \mathcal{C}$ is *additive* if

$$F(M \oplus N) = F(M) \oplus F(N) ,$$

for all objects $M,N \in C$. A functor F is <u>left-exact</u> if for every short exact sequence

$$0 \to M_1 \xrightarrow{f} M_2 \xrightarrow{g} M_3 \to 0 ,$$

the complex

$$0 \longrightarrow FM_1 \xrightarrow{Ff} FM_2 \xrightarrow{Fg} FM_3$$

is exact. A left-exact functor is additive. An example of a left-exact (covariant) functor is $\mathrm{Hom}_A(N,-)$ for some fixed $N \in C$.

2.35 DEFINITION. Take an injective resolution of $M \in C = \underline{A\text{-Mod}}$, say

$$I:\ 0 \longrightarrow M \xrightarrow{\delta_{-1}} I_0 \xrightarrow{\delta_0} I_1 \xrightarrow{\delta_1} \cdots .$$

Let F be a covariant left-exact functor. Applying F to I, we obtain a cochain complex

$$FI:\ 0 \longrightarrow FM \xrightarrow{F\delta_{-1}} FI_0 \xrightarrow{F\delta_0} FI_1 \xrightarrow{F\delta_1} \cdots .$$

The n-th <u>right derived functor</u> R^nF is defined on objects by

$$R^nF(M) = H^n(FI) = \ker F\delta_n / \mathrm{im}\ F\delta_{n-1} .$$

(For $f \in C(M,N)$ there is a natural definition of R^nFf so that R^nF is indeed a (covariant) functor. We won't define R^nFf here, except to note that if $f:M \to M$ is given by multiplication by $x \in A$, then R^nFf is also given by multiplication by x.) Note that by left-exactness $R^0F(M) \cong F(M)$.

One could also define the right derived functor of a contravariant functor, and the left derived functors of covariant and contravariant functors (which would include $\mathrm{Tor}^n_A(-,N)$, $\mathrm{Tor}^n_A(M,-)$, and $\mathrm{Ext}_A(-,N)$ as special cases), but we will only need Definition 2.35 here.

2.36 PROPOSITION. In the setup of the previous definition, let $0 \to M_1 \xrightarrow{f} M_2 \xrightarrow{g} M_3 \to 0$ be a short exact sequence in A-Mod. There is then a long exact sequence

$$\cdots \longrightarrow R^nFM_1 \xrightarrow{R^nf} R^nFM_2 \xrightarrow{R^ng} R^nFM_3 \xrightarrow{\omega_n} R^{n-1}FM_1 \longrightarrow \cdots$$

$$\cdots \longrightarrow R^1FM_3 \xrightarrow{\omega_1} R^0FM_1 \xrightarrow{R^0f} R^0FM_2 \xrightarrow{R^0g} R^0FM_3 \longrightarrow 0 .$$

(We will not define the homomorphisms ω_n here.)

2.37 DEFINITION. An augmented chain complex (over a ring A) is a pair (C,ε), where $C = \{C_q, \partial_q\}$ is a chain complex satisfying $C_q = 0$ if $q < 0$, and $\varepsilon: C_0 \to A$ is an epimorphism satisfying $\varepsilon\partial_1 = 0$ (so that $A \cong C_0/\ker \varepsilon$). This is denoted

$$(C,\varepsilon): \cdots \xrightarrow{\partial_2} C_1 \xrightarrow{\partial_1} C_0 \xrightarrow{\varepsilon} A \longrightarrow 0 .$$

The reduced homology groups $\tilde{H}_q(C)$ (with respect to the augmentation ε) are the homology groups of the augmented chain complex (C,ε). Because A is a projective (in fact, free) A-module, the epimorphism $\varepsilon: C_0 \to A$ splits. It follows that homology and reduced homology are related by

$$H_q(C) \cong \begin{cases} \tilde{H}_q(C), & q > 0 \\ \tilde{H}_0(C) \oplus A, & q = 0 . \end{cases}$$

Similarly of course we can define augmented cochain complex, reduced cochain complex, and reduced cohomology groups. The only difference is that the monomorphism $\varepsilon: A \to C_0$ will not in general split, since A need not be an injective A-module. Of course, if A is a field k, then A is injective and thus $H^0(C) \cong \tilde{H}^0(C) \oplus k$.

2.38 DEFINITION. Let $C = \{C_q, \partial_q\}$ and $C' = \{C'_q, \partial'_q\}$ be two chain (or cochain) complexes over A. The <u>tensor product</u> $C \otimes C' = \{D_n, \varepsilon_n\}$ is the chain (or cochain) complex defined by

$$D_n = (C \otimes C')_n = \coprod_{i+j=n} (C_i \otimes C'_j)$$

$$\varepsilon_n(C_i \otimes C'_j) = \partial_i C_i \otimes C'_j + (-1)^i C_i \otimes \partial'_j C'_j .$$

The reader should check that $\varepsilon_{n-1}\varepsilon_n = 0$ (or $\varepsilon_n \varepsilon_{n-1} = 0$).

For example, in Chapter I.6 there is considered over a ring R a complex $\bigotimes_{i=1}^{s} (0 \to R \to R_{y_i} \to 0)$. When $s = 2$ this becomes

$$0 \to R \otimes R \to (R \otimes R_{y_1}) \oplus (R \otimes R_{y_2}) \to R_{y_1} \otimes R_{y_2} \to 0 .$$

Now $R \otimes M \cong R$ and $R_{y_1} \otimes R_{y_2} \cong R_{y_1 y_2}$. Hence we obtain

$$0 \to R \to R_{y_1} \oplus R_{y_2} \to R_{y_1 y_2} \to 0 .$$

2.39 DEFINITION. Let M be an A-module and $x_1, \cdots, x_r \in A$. For $1 \le i \le r$ let Ae_i be a free A-module of rank one with a specified basis e_i . Let $K(x_i)$ denote the chain complex satisfying:

$$K_0(x_i) = A, \; K_1(x_i) = Ae_i$$

$$K_q(x_i) = 0 \quad \text{if} \quad q \neq 0, 1$$

$$\partial_1(xe_i) = xx_i .$$

This is denoted

$$K(x_i) \colon 0 \to Ae_i \xrightarrow{x_i} A \to 0 .$$

If M is an A-module, then we have a complex

$$K(x_i, M) = K(x_i) \otimes M: 0 \to Me_i \xrightarrow{x_i} M \to 0 .$$

Define the Koszul complex $K(x_1, \cdots, x_r, M)$ with respect to $x_1, \cdots, x_r$ by

$$K(x_1, \cdots, x_r, M) = K(x_1, M) \otimes \cdots \otimes K(x_r, M) .$$

If we put $e_{i_1 \cdots i_q} = u_1 \otimes \cdots \otimes u_r$, where $u_i = e_i$ for $i \in \{i_1, \cdots, i_q\}$ and $u_i = 1$ for other i, then $K_q(x_1, \cdots, x_n, M)$ is a free A-module with basis $\{e_{i_1 \cdots i_q} \mid 1 \leq i_1 < \cdots < i_q \leq r\}$ and thus of rank $\binom{r}{q}$. If $m \in M$, then

$$\partial_q(m\, e_{i_1 \cdots i_q}) = \sum_{j=1}^{q} (-1)^{j-1} x_{i_j}\, m\, e_{i_1 \cdots \hat{i}_j \cdots i_q} ,$$

where $\hat{i}_j$ denotes that i_j is missing. This formula is called a Koszul relation.

§3. Topology.

We now give some basic definitions and results from algebraic topology. Any text on algebraic topology should suffice as a reference; for the most part we follow Spanier.

An (abstract) simplicial complex Δ on a vertex set V is a collection of subsets F of V satisfying:

(a) if $x \in V$ then $\{x\} \in \Delta$,

(b) if $F \in \Delta$ and $G \subset F$, then $G \in \Delta$.

Elements of Δ are called faces or simplices. If $|F| = q+1$, then F is a q-face or q-simplex. We frequently identify the vertex x with the face $\{x\}$.

Suppose V is finite, say $V = \{x_1, \cdots, x_n\}$. Let e_i be the i-th unit coordinate vector in $\mathbb{R}^n$. Given a subset $F \subseteq V$, define

$$|F| = cx\{e_i \mid x_i \in F\} ,$$

where cx denotes convex hull. Thus if F is an (abstract) q-simplex, then $|F|$ is a geometric q-simplex in $\mathbb{R}^n$. Define the geometric realization $|\Delta|$ of the simplicial complex Δ by

$$|\Delta| = \bigcup_{F\in\Delta} |F| .$$

Thus $|\Delta|$ inherits from the usual topology on $\mathbb{R}^n$ the structure of a topological space. If X is a topological space homeomorphic to $|\Delta|$, then we (somewhat inaccurately) call Δ a triangulation of X.

An oriented q-simplex of Δ is a q-simplex F together with an equivalence class of total orderings of F, two orderings being equivalent if they differ by an even permutation of the vertices. Denote by $[v_0, v_1,\cdots, v_q]$ the oriented q-simplex consisting of the q-simplex $F = \{v_0, v_1,\cdots, v_q\}$, together with the equivalence class of orderings containing $v_0 < v_1 <\cdots< v_q$. Fix a ring A (commutative with 1). Let $C_q(\Delta)$ be the free A-module with basis consisting of the oriented q-simplices in Δ, modulo the relations $\sigma_1 + \sigma_2 = 0$ whenever σ_1 and σ_2 are different oriented q-simplices corresponding to the same q-simplex of Δ. Thus $C_q(\Delta) = 0$ for $q < 0$, and for $q \geq 0$ $C_q(\Delta)$ is a free A-module with rank equal to the number of q-simplices of Δ. If Δ is empty, then $C_q(\Delta) = 0$ for all q.

We define homomorphisms $\partial_q : C_q(\Delta) \to C_{q-1}(\Delta)$ for $q \geq 1$ by defining them on the basis elements by

$$\partial_q[v_0, v_1,\cdots, v_q] = \sum_{i=0}^{q} (-1)^i [v_0, v_1,\cdots, \hat{v}_i,\cdots, v_q] ,$$

where $\hat{v}_i$ denotes that v_i is missing. It is easily verified that ∂_q indeed extends to a homomorphism $C_q(\Delta) \to C_{q-1}(\Delta)$, and that $\partial_q\partial_{q+1} = 0$. The chain complex $C(\Delta) = \{C_q(\Delta), \partial_q\}$ is the oriented chain complex of Δ. Define an augmentation $\varepsilon: C_0(\Delta) \to A$ by $\varepsilon(x) = 1$ for every vertex $x \in V$. The augmented chain complex $(C(\Delta),\varepsilon)$ is the augmented oriented chain complex of Δ (over A).

3.1 DEFINITION. The <u>q-th reduced homology group of Δ with coefficients A</u>, denoted $\tilde{H}_q(\Delta;A)$, is defined to be the q-th homology group of the augmented oriented chain complex of Δ (over A).

3.2 DEFINITION. The <u>reduced Euler characteristic</u> $\tilde{\chi}(\Delta)$ of Δ is defined by

$$\tilde{\chi}(\Delta) = \sum_{q\geq -1} (-1)^q \operatorname{rank} \tilde{H}_q(\Delta;A) .$$

It is independent of A and is also given by

$$\tilde{\chi}(\Delta) = -1 + f_0 - f_1 + \cdots,$$

where f_q is the number of q-simplices in Δ. If $\chi(\Delta)$ is the ordinary Euler characteristic then $\tilde{\chi}(\Delta) = \chi(\Delta) - 1$.

<u>Note</u>: If $\Delta \neq \phi$, then $\tilde{H}_q(\Delta;A) = 0$ for $q < 0$. If $\Delta = \phi$, then

$$\tilde{H}_q(\phi;A) \cong \begin{cases} A, & q=-1 \\ 0, & q\neq -1 \end{cases}$$

In particular, $\tilde{\chi}(\phi) = -1$.

3.3 PROPOSITION. If $\Delta \neq \phi$, then $\tilde{H}_0(\Delta;A)$ is a free A-module whose rank is one less than the number of connected components of Δ.

We now wish to define the homology groups of a space X, rather than a simplicial complex Δ. Let X be a topological space. Let Δ^q denote the standard q-dimensional ordered geometric simplex $\langle p_0, \cdots, p_q\rangle$ whose vertices p_i are the unit coordinate vectors in $\mathbb{R}^{q+1}$. A <u>singular q-simplex</u> in X is a continuous map

$$\sigma: \Delta^q \to X .$$

Let $C_q(X)$ be the free A-module generated by all singular q-simplices. The elements of C_q are formal finite linear combinations $\sum_\sigma c_\sigma \sigma$, where σ is a singular q-simplex and $c_\sigma \in A$. Given a vertex p_i of Δ_q, there is an obvious linear map $e^i_q\colon \Delta^{q-1} \to \Delta^q$ which sends Δ^{q-1} to the face of Δ^q opposite p_i. The <u>i-th face</u> of σ, denoted by $\sigma^{(i)}$, is defined to be the singular (q-1)-simplex which is the composite

$$\sigma^{(i)} = \sigma \circ e^i_q\colon \Delta^{q-1} \to \Delta^q \to X .$$

We now define a linear map (= A-module homomorphism) $\partial_q\colon C_q \to C_{q-1}$ by

$$\partial_q(\sigma) = \sum_{i=0}^{q} (-1)^i \sigma^{(i)} ,$$

where σ is a singular q-simplex. It is easily checked that $\partial_{q-1}\partial_q = 0$, so $C(X) = \{C_q(X), \partial_q\}$ is a chain complex, the <u>singular chain complex</u> of X (over A). Define an augmentation $\varepsilon\colon C_0(X) \to A$ by $\varepsilon(\sigma) = 1$ for all singular 0-simplices σ. The augmented chain complex $\tilde{C}(X)$ is the <u>augmented singular chain complex</u> of X (over A).

3.4 DEFINITION. The <u>q-th reduced singular homology group of X with coefficients A</u>, denoted $\tilde{H}_q(X;A)$, is the q-th homology group of the augmented singular chain complex of X (over A).

3.5 DEFINITION. The <u>reduced Euler characteristic</u> $\tilde{\chi}(X)$ of X is defined by

$$\tilde{\chi}(X) = \sum_{q \geq -1} (-1)^q \operatorname{rank} \tilde{H}_q(X;A) .$$

It is independent of A.

If Δ is a simplicial complex and Δ_1 and Δ_2 are subcomplexes of Δ, then there is an exact sequence (whose definition we omit)

$$\cdots \to \tilde{H}_q(\Delta_1 \cap \Delta_2) \to \tilde{H}_q(\Delta_1) \oplus \tilde{H}_q(\Delta_2) \to$$

$$\tilde{H}_q(\Delta_1 \cup \Delta_2) \to \tilde{H}_{q-1}(\Delta_1 \cap \Delta_2) \to \cdots$$

(with all coefficients A), called the reduced Mayer-Vietoris sequence of Δ_1 and Δ_2. Similarly, if X is a topological space and X_1, X_2 are "nice" subspaces (e.g., if $X_1 \cup X_2 = (\text{int}_{X_1 \cup X_2} X_1) \cup (\text{int}_{X_1 \cup X_2} X_2)$, where $\text{int}_Y Z$ denotes the relative interior of Z in the space Y), then we have a reduced Mayer-Vietoris sequence of X_1 and X_2 exactly analogous to that of Δ_1 and Δ_2.

We now come to the relationship between simplicial and singular homology.

3.6 THEOREM. Let Δ be a finite simplicial complex and $X = |\Delta|$. Then there is a (canonical) isomorphism

$$\tilde{H}_q(\Delta;A) \cong \tilde{H}_q(X;A) ,$$

for all q.

3.7 PROPOSITION. Let S^{d-1} denote a $(d-1)$-dimensional sphere. Then

$$\tilde{H}_q(\Delta;A) \cong \begin{cases} A, & q = d-1 \\ 0, & q \neq d-1 \end{cases} .$$

3.8 DEFINITION. A simplicial complex Δ or topological space X is acyclic (over A) if its reduced homology with coefficients A vanishes in all degrees q. (Thus the null set is not acyclic, since $\tilde{H}_{-1}(\phi;A) \cong A$.)

3.9 DEFINITION. Let Y be a subspace of X. Then the singular chain module $C_q(Y)$ is a submodule of $C_q(X)$, so we have a quotient complex $C(X,Y) = C(X)/C(Y) = \{C_q(X)/C_q(Y), \bar{\partial}_q\}$. Define the relative homology of X modulo Y (with coefficients A) by

$$H_q(X,Y;A) = H_q(C(X,Y)) .$$

We next want to define reduced cohomology of simplicial complexes and spaces. The simplest way (though not the most geometric) is to dualize the corresponding chain complexes.

3.10 DEFINITION. Let $C'(\Delta) = C(\Delta,\varepsilon)$ be the augmented oriented chain complex of the simplicial complex Δ, over the ring A. The q-th reduced singular cohomology group of Δ with coefficients A is defined to be

$$\tilde{H}^q(\Delta;A) = \tilde{H}^q(\mathrm{Hom}_A(C'(\Delta),A)) ,$$

where $\mathrm{Hom}_A(C'(\Delta),A)$ is the cochain complex obtained by applying the functor $\mathrm{Hom}_A(-,A)$ to $C'(\Delta)$. Exactly analogously define $\tilde{H}^q(X;A)$ and $H^q(X,Y;A)$. Sometimes one identifies the free modules $C_q(\Delta)$ and $C^q(\Delta) = \mathrm{Hom}_A(C_q(\Delta),A)$ by identifying the basis of oriented q-chains σ of $C_q(\Delta)$ with its dual basis in $C^q(\Delta)$. Similarly one can identify $C_q(X)$ with $C^q(X)$.

There is a close connection between homology and cohomology of Δ or X arising from the "universal-coefficient theorem for cohomology." We merely mention the (easy) special case that when A is a field k, there are "canonical" isomorphisms

$$\tilde{H}_q(\Delta;k) \xrightarrow{\cong} \mathrm{Hom}_k(\tilde{H}^q(\Delta;k),\ k)$$

$$\tilde{H}_q(X;k) \xrightarrow{\cong} \mathrm{Hom}_k(\tilde{H}^q(X;k),\ k) .$$

Thus in particular when $\tilde{H}_q(\Delta;k)$ is finite-dimensional (e.g., when Δ is finite), we have $\tilde{H}_q(\Delta;k) \cong \tilde{H}^q(\Delta;k)$ and similarly for X, but these isomorphisms are not canonical.

3.11 DEFINITION. A topological n-manifold (without boundary) is a Hausdorff space in which each point has an open neighborhood homeomorphic to $\mathbb{R}^n$. An n-manifold with boundary is a Hausdorff space X in which each point has an open neighborhood which is homeomorphic with $\mathbb{R}^n$ or $\mathbb{R}^n_+ = \{(x_1,\cdots, x_n) \in \mathbb{R}^n \mid x_i \geq 0\}$. The boundary ∂X of X consists of those points with no open neighborhood homeomorphic to $\mathbb{R}^n$. It follows easily that ∂X is either void or an (n-1)-manifold.

Suppose X is a compact connected n-manifold with boundary. Then one can show $H_n(X,\partial X;A)$ is either void or isomorphic to A.

3.12 DEFINITION. A compact connected n-manifold X with boundary is orientable (over A) if $H_n(X,\partial X;A) = A$. (The usual definition of orientable is more technical but equivalent to the one given here.)

3.13 PROPOSITION. Every compact connected n-manifold with boundary is orientable over a field of characteristic two.

3.14 POINCARÉ DUALITY THEOREM. If a compact connected n-manifold X is orientable over A, then $H_q(X;A) \cong H^{n-q}(X;A)$.

3.15 DEFINITION. An n-dimensional pseudomanifold without boundary (resp., with boundary) is a simplicial complex Δ such that:

(a) Every simplex of Δ is the face of an n-simplex of Δ.

(b) Every (n-1)-simplex of Δ is the face of exactly two (resp., at most two) n-simplices of Δ.

(c) If F and F' are n-simplices of Δ, there is a finite sequence $F = F_1, F_2, \dots, F_m = F'$ of n-simplices of Δ such that F_i and F_{i+1} have an (n-1)-face in common for $1 \le i < m$.

The boundary $\partial\Delta$ of a pseudomanifold Δ consists of those faces F contained in some (n-1)-simplex of Δ which is the face of exactly one n-simplex of Δ.

3.16 PROPOSITION and DEFINITION. Let Δ be a finite n-dimensional pseudomanifold with boundary. Then either $H_n(\Delta,\partial\Delta;A) \cong A$ or 0. In the former case we say that Δ is orientable over A; otherwise nonorientable.

3.17 DEFINITION. Let I be the unit interval $[0,1]$. The suspension ΣX of a topological space X is defined to be the quotient space of $X \times I$ in which $X \times 0$ is identified to one point and $X \times 1$ is identified to another point. The n-fold suspension $\Sigma^n X$ is defined recursively by $\Sigma^n X = \Sigma(\Sigma^{n-1} X)$.

3.18 PROPOSITION. For any X and q,

$$\tilde{H}_q(X;A) \cong \tilde{H}_{q+1}(\Sigma X;A) .$$

CHAPTER I

NONNEGATIVE INTEGRAL SOLUTIONS TO LINEAR EQUATIONS

§1. Integer stochastic matrices (magic squares)

The first topic will concern the problem of solving linear equations in nonnegative integers. In particular, we will consider the following problem which goes back to MacMahon. Let

$$H_n(r) := \text{number of } n \times n \ \mathbb{N}\text{-matrices having line sums } r,$$

where a line is a row or column, and an $\mathbb{N}$-matrix is a matrix whose entries belong to $\mathbb{N}$. Such a matrix is called an integer stochastic matrix or magic square. Keeping r fixed, one finds that $H_n(0) = 1$, $H_n(1) = n!$, and Anand, Dumir and Gupta [A-D-G] showed that

$$\sum_{n \geq 0} \frac{H_n(2)x^n}{(n!)^2} = \frac{e^{x/2}}{\sqrt{1-x}} .$$

See also Stanley [St_5, Ex. 6.11]. Keeping n fixed, one finds that $H_1(r) = 1$, $H_2(r) = r + 1$, and MacMahon [MM, Sect. 407] showed that

$$H_3(r) = \binom{r+4}{4} + \binom{r+3}{4} + \binom{r+2}{4} .$$

Guided by this evidence Anand, Dumir and Gupta [A-D-G] formulated the following

1.1 CONJECTURE. Fix $n \geq 1$. Then

(i) $H\ (r) \in \mathbb{Q}[r]$

(ii) $\deg H_n = (n-1)^2$

(iii) $H_n(-1) = H_n(-2) = \cdots = H_n(-n+1) = 0,$

$$H_n(-n-r) = (-1)^{n-1} H_n(r).$$

This conjecture can be shown equivalent to:

$$\sum_{r\geq 0} H_n(r)\lambda^r = \frac{h_0 + h_1\lambda + \cdots + h_d\lambda^d}{(1-\lambda)^{(n-1)^2+1}} \quad , \quad d = n^2 - 3n + 2 \quad ,$$

$h_0 + h_1 + \cdots + h_d \neq 0$, and $h_i = h_{d-i}$, $i = 0,1,\ldots,d$.

The following additional conjectures can be made:

(iv) $h_i \geq 0$,

(v) $h_0 \leq h_1 \leq \ldots \leq h_{[d/2]}$.

We will verify conjectures (i) to (iv). Conjecture (v) is still open. The solution will appear as a special case of solving linear diophantine equations. This will be done in a ring-theoretic setting, and we will now review the relevant commutative algebra.

§2. Graded algebras and modules.

Let k be a field, and let R be an $\mathbb{N}^m$-graded connected commutative k-algebra with identity. Thus,

$$R = \coprod_{\alpha\in\mathbb{N}^m} R_\alpha \text{ (vector space direct sum)}, \quad R_\alpha R_\beta \subseteq R_{\alpha+\beta} \ , \ R_0 = k.$$

Elements $x \in R_\alpha$ are said to be homogeneous of degree α, $\deg x = \alpha$. Let

$$\mathcal{H}(R) = \bigcup_\alpha R_\alpha \quad ,$$

and

$$R_+ = \coprod_{\alpha \neq 0} R_\alpha .$$

The ideal R_+ (called the <u>irrelevant ideal</u>) is maximal; in fact, it is the unique maximal homogeneous ideal.

A $\mathbb{Z}^m$-graded R-module M has a decomposition of the form

$$M = \coprod_{\alpha \in \mathbb{Z}^m} M_\alpha \text{ (vector space direct sum)}, \; R_\alpha M_\beta \subseteq M_{\alpha+\beta}$$

and a map $\phi : M \to N$ between two such modules is <u>degree-preserving</u> (or <u>graded</u>) if $\phi(M_\alpha) \subseteq N_\alpha$ for all α. As above, write $H(M) = \bigcup_\alpha M_\alpha$. A submodule $N \subseteq M$ is <u>homogeneous</u> if generated by homogeneous elements, and such a submodule has the structure of a graded R-module by $N_\alpha = N \cap M_\alpha$. In particular, this defines <u>homogeneous ideals</u>. If N is a homogeneous submodule of M, then also M/N is a graded R-module:

$$M/N = \coprod_\alpha (M/N)_\alpha \quad , \quad (M/N)_\alpha = M_\alpha / N_\alpha \quad .$$

From now on we assume that R is a finitely generated k-algebra, i.e., of <u>finite type</u> (equivalently, noetherian), and that all graded R-modules are finitely-generated, unless the contrary is explicitly stated. Sometimes it is convenient to be able to consider R-modules as modules over a polynomial ring. Let $y_1, y_2, \ldots, y_s \in H(R_+)$, $\deg y_i = \delta_i \in \mathbb{N}^m - \{0\}$. Introduce new variables Y_i with $\deg Y_i = \delta_i$, and let $A = k[Y_1, \ldots, Y_s]$. Define a graded A-module structure on R (and hence on any graded R-module) by $Y_i \cdot x = y_i x$ if $x \in R$. R will be a finitely-generated A-module if and only if R is integral over the subring $k[y_1, \ldots, y_s]$. In particular, if $y_1, \ldots, y_s$ generate R then the homomorphism $A \to R$ defined by $Y_i \mapsto y_i$ is surjective, so $R = A/I$ for some homogeneous ideal $I \subseteq A$.

Let M be a finitely-generated $\mathbb{Z}^m$-graded R-module.

2.1 DEFINITION. $H(M,\alpha) = \dim_k M_\alpha < \infty$, $\alpha \in \mathbb{Z}^m$,

$$F(M,\lambda) = \sum_{\alpha \in \mathbb{Z}^m} H(M,\alpha) \lambda^\alpha .$$

$H(M,\alpha)$ is called the Hilbert function of M and $F(M,\lambda)$ is called the Hilbert series of M. Here $\lambda = (\lambda_1,\lambda_2,\ldots,\lambda_m)$, $\alpha = (\alpha_1,\alpha_2,\ldots,\alpha_m)$, and $\lambda^\alpha = \lambda_1^{\alpha_1}\lambda_2^{\alpha_2}\cdots\lambda_m^{\alpha_m}$. Clearly, $F(M,\lambda) \in \mathbb{Z}[[\lambda_1,\ldots,\lambda_m]][\lambda_1^{-1},\ldots,\lambda_m^{-1}]$, since there cannot be arbitrarily large negative exponents due to finite generation.

Given M and $\theta \in R$ let $(0 : \theta) := \{u \in M \mid \theta u = 0\}$. $(0 : \theta)$ is a homogeneous submodule of M. Proofs for the following lemma and theorem can be found in Atiyah-Macdonald [A-M, ch. 11] or in Stanley [St_6, Thm. 3.1].

2.2 LEMMA. Let $\theta \in R_\alpha$, $\alpha \neq 0$. Then

$$F(M,\lambda) = \frac{F(M/\theta M,\lambda) - \lambda^\alpha F((0 : \theta),\lambda)}{1 - \lambda^\alpha}$$

This lemma easily implies the following:

2.3 THEOREM. Suppose R is generated by $y_1, y_2,\ldots,y_s$, $\deg y_i = \delta_i \neq 0$. Then for some $\beta \in \mathbb{Z}^m$ and $P(M,\lambda) \in \mathbb{Z}[\lambda]$,

$$F(M,\lambda) = \lambda^\beta \frac{P(M,\lambda)}{\prod_{i=1}^{s}(1 - \lambda^{\delta_i})}.$$

§3. Elementary aspects of $\mathbb{N}$-solutions to linear equations.

Let us now return to consider $\mathbb{N}$-solutions to linear systems of equations over $\mathbb{Z}$. Many of the details which we omit on this topic may be found in [St_{11}]. Let Φ be an $r \times n$ $\mathbb{Z}$-matrix, $r \leq n$, and rank $\Phi = r$. Let $E_\Phi := \{\beta \in \mathbb{N}^n \mid \Phi\beta = 0\}$, and for $\alpha \in \mathbb{Z}^r$ let $E_{\Phi,\alpha} := \{\beta \in \mathbb{N}^n \mid \Phi\beta = \alpha\}$. E_Φ is clearly a submonoid of $\mathbb{N}^n$, and $E_{\Phi,\alpha}$ is an "E -module", i.e., $E_\Phi + E_{\Phi,\alpha} \subseteq E_{\Phi,\alpha}$.

Let $R_\Phi := kE_\Phi$, the monoid algebra of E_Φ over k. We identify $\beta \in E_\Phi$ with $x^\beta = x_1^{\beta_1}x_2^{\beta_2}\cdots x_n^{\beta_n}$, so that $R_\Phi \subseteq k[x_1,x_2,\ldots,x_n]$ as a subalgebra generated by monomials. Let $M_{\Phi,\alpha} := kE_{\Phi,\alpha}$. Then $M_{\Phi,\alpha}$ is a $\mathbb{Z}^n$-graded R_Φ-module, with grading $\deg x^\beta = \beta$. Clearly,

$$F(R_\Phi,\lambda) = \sum_{\beta\in E_\Phi} \lambda^\beta ,$$

and also

$$F(M_{\Phi,\alpha},\lambda) = \sum_{\beta\in E_{\Phi,\alpha}} \lambda^\beta .$$

Hence, in this case the ring and module are completely determined by their Hilbert series.

An aside on invariant theory:

Let γ_i denote the i-th column of Φ, so $\Phi = [\gamma_1,\gamma_2,\dots,\gamma_n]$, and let

$$T = \left\{ \begin{bmatrix} u^{\gamma_1} & & & 0 \\ & u^{\gamma_2} & & \\ & & \ddots & \\ 0 & & & u^{\gamma_n} \end{bmatrix} \quad : \quad u = (u_1,u_2,\dots,u_r) \in (k^*)^r \right\} .$$

T is an r-dimensional torus $\subseteq GL(n,k)$, considered as an algebraic group over k, and T acts on $R = k[x_1,x_2,\dots,x_n]$ by $\tau x_i = u^{\gamma_i}x_i$, where $\tau = \mathrm{diag}(u^{\gamma_1},u^{\gamma_2},\dots,u^{\gamma_n})$. Then

$$R^T := \{t \in R \mid \tau \cdot t = t , \forall \tau \in T\} \cong R_\Phi .$$

Also, for $\alpha \in \mathbb{Z}^r$ and $\tau = \mathrm{diag}(u^{\gamma_1},u^{\gamma_2},\dots,u^{\gamma_n}) \in T$, let $\chi_\alpha(\tau) = u^\alpha \in k^*$. χ_α is a one-dimensional character of T and

$$R^T_{\chi_\alpha} := \{t \in R \mid \tau \cdot t = \chi_\alpha(\tau) \cdot t , \quad \forall \tau \in T\} \cong M_{\Phi,\alpha} .$$

Thus, some of the developments we present for R_Φ (and $M_{\Phi,\alpha}$) can be seen as special cases of general results about invariant rings of reductive algebraic groups acting on polynomial rings (e.g. Cohen-Macaulayness, due to Hochster).

3.1 THEOREM. R_Φ is a finitely-generated k-algebra.

Proof. Let I be the ideal of $R = k[x_1,x_2,\ldots,x_n]$ generated by $(R_\Phi)_+$. By the Hilbert Basis Theorem I is finitely-generated, i.e., we can find $x^{\delta_1}, x^{\delta_2}, \ldots, x^{\delta_t}$ in $(R_\Phi)_+$ which generate I as an ideal of R.

We want to show that E_Φ is a finitely-generated monoid. Claim: $\delta_1,\delta_2,\ldots,\delta_t$ generate E_Φ. Let $\beta \in E_\Phi$. Since $x^\beta \in I$, we get $\beta = \delta_i + \gamma$, $\gamma \in \mathbb{N}^m$. But $\beta,\delta_i \in E_\Phi$ implies $\gamma \in E_\Phi$ (this is the crucial property of this monoid). Having peeled off one generator δ_i, we continue until we get β expressed as a sum of δ_i's. □

3.2 THEOREM. $M_{\Phi,\alpha}$ is a finitely-generated R_Φ-module.

The proof is similar.

We want now to find a "smallest" subset $\{\delta_1,\delta_2,\ldots,\delta_t\} \subseteq E_\Phi$ such that R_Φ is a finitely-generated $k[x^{\delta_1}, x^{\delta_2}, \ldots, x^{\delta_t}]$-module.

3.3 DEFINITION. $\beta \in E_\Phi$ is fundamental if $\beta = \gamma + \delta$, $\gamma,\delta \in E_\Phi$ implies $\gamma = \beta$ or $\delta = \beta$.

FUND_Φ := set of fundamental elements of E_Φ.

It is clear that FUND_Φ generates E_Φ, and that every set which generates E_Φ contains FUND_Φ. In particular, $|\mathrm{FUND}_\Phi| < \infty$ and $R_\Phi = k[x^\delta \mid \delta \in \mathrm{FUND}_\Phi]$.

3.4 DEFINITION. $\beta \in E_\Phi$ is completely fundamental if whenever $n > 0$ and $n\beta = \gamma + \delta$ for $\gamma,\delta \in E_\Phi$, then $\gamma = n_1\beta$ for some $0 \le n_1 \le n$.

CF_Φ : = set of completely fundamental elements of E_Φ.

3.5 EXAMPLE. Let $\Phi = [1 \;\; 1 \;\; -2]$, so we are looking for $\mathbb{N}$-solutions to $x + y - 2z = 0$. Then

$\mathrm{FUND}_\Phi = \{(201), (021), (111)\}$, and

$CF_\Phi = \{(201), (021)\}$, since $2(111) = (201) + (021)$.

In the general situation, consider now the set of $\mathbb{R}^+$-solutions β to $\Phi\beta = 0$. It forms a convex polyhedral cone C_Φ whose unique vertex is the origin. The integer points nearest 0 on each extreme ray of C_Φ form the set CF_Φ. Furthermore, the faces of C_Φ (intersections with supporting hyperplanes) are in one-to-one correspondence with $\{\text{supp } \beta | \beta \in E_\Phi\}$, where $\beta = (\beta_1, \beta_2, \ldots, \beta_n) \in \mathbb{N}^n$, $\text{supp } \beta = \{i | \beta_i > 0\}$.

3.6 EXAMPLE. Consider $x_1 + x_2 - x_3 - x_4 = 0$. The cone of solutions looks like

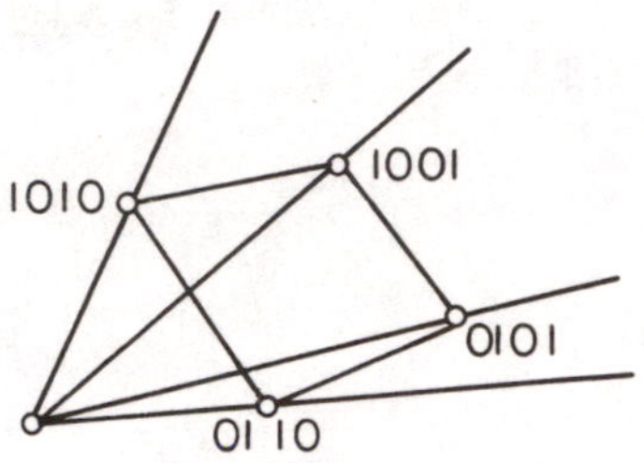

so it has the facial structure of a square. Also, the supports of solutions, ordered by inclusion, yield the face-lattice of a square:

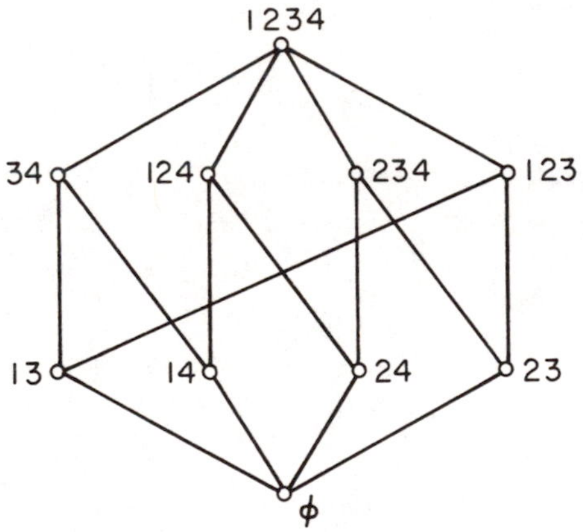

3.7 THEOREM. Let $\delta_1, \delta_2, \ldots, \delta_t \in E_\Phi$, and $S = k[x^{\delta_1}, x^{\delta_2}, \ldots, x^{\delta_t}] \subseteq R_\Phi$. Then, R_Φ is a finitely-generated S-module (equivalently: integral over S) if and only if for every $\beta \in CF_\Phi$ there are $1 \le i \le t$ and $j > 0$ such that $\delta_i = j\beta$.

Proof. See Prop. 2.2 in Stanley [St_1]. Visualizing the situation geometrically a proof can be gleaned from the following remarks.

"$\Leftarrow$": If $\delta \in E_\Phi$, then $n\delta = \sum_{\beta \in CF_\Phi} a_\beta \beta$, which in R_Φ is $(x^\delta)^n = \Pi (x^\beta)^{a_\beta}$. Hence R_Φ is integral over $k[x^\beta | \beta \in CF_\Phi]$.

"$\Rightarrow$": If the δ_i's miss some extreme ray $j\beta$, there is no way of reaching the entire ray using nonnegative linear combinations. □

3.7 COROLLARY. When $F(R_\Phi, \lambda) = \sum_{\beta \in E_\Phi} \lambda^\beta$ is written in lowest terms, the denominator is $\prod_{\beta \in CF_\Phi} (1 - \lambda^\beta)$.

Proof. Let $CF_\Phi = \{\beta_1, \beta_2, \ldots, \beta_s\}$, and let $A = k[Y_1, Y_2, \ldots, Y_s]$, $\deg Y_i = \beta_i$. R_Φ is an A-module via the multiplication $Y_i \cdot x^\delta = x^{\delta + \beta_i}$. In fact, the theorem shows that R_Φ is a finitely-generated A-module. Hence $\prod_{\beta \in CF_\Phi} (1 - \lambda^\beta)$ is a denominator. To see that it is the least denominator, consult [St_1, Thm. 2.5]. □

In an analogous way it can be shown that either $F(M_{\Phi,\alpha}, \lambda) = 0$ or else has the same denominator as $F(R_\Phi, \lambda)$. The former case occurs, e.g., for $\Phi = [2\ -2]$, $\alpha = 1$.

§4. Integer stochastic matrices again.

As an application of the preceding, consider

$$E_\Phi = \{(a_{ij})_{i,j=1}^n | a_{ij} \in \mathbb{N}, \text{ line sums equal}\} ,$$

$$R_\Phi = kE_\Phi = k\left[\prod_{i,j=1}^n x_{ij}^{a_{ij}}\right] .$$

The question of which are the completely fundamental elements of E_Φ is in this case answered by the Birkhoff - Von Neumann theorem (which says that the extreme points of the convex set of doubly stochastic matrices are the permutation matrices). It follows that every

$\alpha \in E_\Phi$ is a sum of permutation matrices, so

$$\mathrm{FUND}_\Phi = \mathrm{CF}_\Phi = \{\text{permutation matrices}\} .$$

Therefore, $F(R_\Phi,\lambda) = \dfrac{P(\lambda)}{\Pi(1-\lambda^\beta)}$, the product taken over all $n \times n$ permutation matrices. Letting $\lambda_{ij} = \begin{cases} x, & i=1 \\ 0, & i\neq 1 \end{cases}$, so that $\Pi\lambda_{ij}^{a_{ij}} = x^{\text{line sum}}$, we derive

$$\sum_{r\geq 0} H_n(r)x^r = \frac{P_1(x)}{(1-x)^{n!}} ,$$

which proves that $H_n(r)$ is a polynomial for large r (cf. the first part of the Anand-Dumir-Gupta conjecture). It will be shown that $H_n(r)$ is a polynomial for <u>all</u> r, i.e., that $\deg P_1 < n!$. The substitution $\lambda_{ij} = \begin{cases} x, & i=1 \\ 0, & i\neq 1 \end{cases}$ is equivalent to "specializing" the $\mathbb{N}^{n^2}$-grading of R_Φ to an $\mathbb{N}$-grading by defining $\deg(\Pi\lambda_{ij}^{a_{ij}})$ = line sum of $(a_{ij}) = \sum_j a_{ij}$.

Now let E_Φ := {n × n <u>symmetric</u> $\mathbb{N}$-matrices having equal line sums}. In this case FUND_Φ is much harder to describe; fundamental elements with arbitrarily large line-sums can be shown to exist.

4.1 THEOREM. If $\beta \in \mathrm{CF}_\Phi$, then the line sum of β is 1 or 2.

<u>Proof</u>. For $\beta \in \mathrm{CF}_\Phi$, by the Birkhoff-Von Neumann theorem $\beta = \sum a_i\pi_i$, π_i permutation matrices, $a_i \in \mathbb{N}$. Hence, $2\beta = \beta + \beta^t = \sum a_i(\pi_i + \pi_i^t)$, from which the proof follows. □

<u>Remark</u>. It is possible to characterize the n × n completely fundamental symmetric matrices, and for $f(n) = |\mathrm{CF}_\Phi|$ it can be shown that

$$\sum_{n\geq 0} \frac{f(n)x^n}{n!} = \left(\frac{1+x}{1-x}\right)^{\frac{1}{4}} e^{\frac{1}{2}x+\frac{1}{2}x^2} .$$

4.2 COROLLARY. Let $S_n(r)$:= # $\{\beta \in E_\Phi \mid \text{line sum of } \beta \text{ is } r\}$. Then

$$\sum_{r\geq 0} S_n(r)x^r = \frac{V(x)}{(1-x)^a(1-x^2)^b} , \quad V(x) \in \mathbb{Z}[x] , \quad a,b \geq 0 .$$

Proof. Specialize the grading: deg x^β = line sum of β and apply Corollary 3.7. □

The form of the generating function reveals that

$$S_n(r) = P_n(r) + (-1)^r Q_n(r) \ , \quad P_n, Q_n \in \mathbb{Q}[r] \quad ,$$

for large r, and since $S_n(r) > 0$ for such r, deg $P_n \geq$ deg Q_n.

§5. Dimension, depth, and Cohen-Macaulay modules.

Now some more review of commutative algebra. Let R be an $\mathbb{N}^m$-graded k-algebra, and let M be a $\mathbb{Z}^m$-graded R-module. The Krull dimension is defined by the following equal numbers:

dim R = maximum number of (homogeneous) elements of R algebraically independent over k
= length of longest chain of prime ideals of R
= order to which $\lambda = 1$ is a pole of $F(R,\lambda)$ (if m = 1).

dim M = dim (R/Ann M)
= order to which $\lambda = 1$ is a pole of $F(M,\lambda)$ (if m = 1).

Although the last condition refers to the m = 1 case, by specialization of the grading it can apply also in the general case.

5.1 DEFINITION. A partial hsop (homogeneous system of parameters) for M is a sequence $\theta_1,\theta_2,\ldots,\theta_r \in H(R_+)$ such that $\dim M/(\theta_1 M + \theta_2 M + \cdots + \theta_r M) = \dim M - r$. A hsop is a partial hsop with r = dim M. Equivalently, $\theta_1,\theta_2,\ldots,\theta_d \in H(R_+)$ is a hsop for M $\Longleftrightarrow$ d = dim M and M is a finitely-generated $k[\theta_1,\theta_2,\ldots,\theta_d]$-module.

Being a partial hsop implies being algebraically independent over k, but not conversely. Under what conditions does an hsop exist?

5.2 NOETHER NORMALIZATION LEMMA. If m = 1 there exists an hsop for M. If $|k| = \infty$ and R is generated by R_1, then we can choose an hsop from R_1.

The theorem fails for m > 1, e.g., R = k[x,y]/(xy) with deg x = (1,0), deg y = (0,1) lacks an hsop.

Let us return momentarily to the ring R_Φ. Using that monomials $x^{\beta_1}, x^{\beta_2}, \ldots, x^{\beta_t}$, $\beta_i \in E_\Phi$, are algebraically independent over k if and only if the vectors $\beta_1, \beta_2, \ldots, \beta_t$ are linearly independent over $\mathbb{Q}$, we find:

$$\begin{aligned}\dim R_\Phi &= \text{maximum number of linearly independent elements in } E_\Phi \\ &= \text{dimension of vector space spanned by } E_\Phi \text{ over } \mathbb{Q} \\ &= n-r \text{ (assuming that } \exists \beta > 0,\ \beta \in E_\Phi).\end{aligned}$$

Now recall the polynomials (for r large) $H_n(r)$ and $P_n(r)$ related to the enumeration of $n \times n$ (symmetric) $\mathbb{N}$-matrices with constant line-sum r.

5.3 COROLLARY. (i) $\deg H_n(r) = (n-1)^2$,

(ii) $\deg P_n(r) = \binom{n}{2}$.

Proof. The degree of these polynomials is one less than the order to which $\lambda = 1$ is a pole of $F(R_\Phi, \lambda)$. In the first case Φ is an $(2n-2) \times n^2$-matrix and in the second case an $(n-1) \times \binom{n+1}{2}$-matrix, so the proof follows from the above description of $\dim R_\Phi$. □

Recall that we had $S_n(r) = P_n(r) + (-1)^r Q_n(r)$, with $\deg Q_n \le \deg P_n = \binom{n}{2}$. Concerning the problem of finding $\deg Q_n$ the following can be proved.

5.4 THEOREM. Let $E_\Phi := \{n \times n$ symmetric $\mathbb{N}$-matrices with equal line sums$\}$, $\deg x^\beta :=$ line sum of β. Let

$$f_n := \min \{k \mid \theta_1, \theta_2, \ldots, \theta_d \text{ is an hsop for } R_\Phi ,$$

$$\deg \theta_1 = \cdots = \deg \theta_k = 2 , \quad \deg \theta_{k+1} = \cdots = \deg \theta_d = 1\} .$$

Then

$$f_n = \begin{cases} \binom{n-1}{2} , & n \text{ odd} \\ \binom{n-2}{2} , & n \text{ even} \end{cases} .$$

5.5 THEOREM. $\deg Q_n \le f_n - 1$ (Conjecture: Equality holds)

Proof. See Stanley [St_7, Prop. 5.4]. □

5.6 DEFINITION. $\theta_1,\theta_2,\ldots,\theta_r \in H(R_+)$ is a homogeneous M-sequence (regular sequence) if θ_{i+1} is a non-zero-divisor on $M/(\theta_1 M + \cdots + \theta_i M)$, $0 \le i < r$. Equivalently, $\theta_1,\theta_2,\ldots,\theta_r$ are algebraically independent over k and M is a free $k[\theta_1,\theta_2,\ldots,\theta_r]$-module.

An M-sequence is a partial hsop, and if m = 1 any two maximal M-sequences have the same length. The latter is not true for m > 1, e.g. letting $R = k[x,y,z]/(xy-z^2)$, deg x = (2,0), deg y = (0,2), and deg z = (1,1), then {x,y} and {z} are maximal homogeneous R-sequences. In terms of the Hilbert series we get the following characterization: $\theta_1,\theta_2,\ldots,\theta_r \in H(R_+)$ is an M-sequence if and only if

$$F(M,\lambda) = \frac{F\left(M/(\theta_1 M + \cdots + \theta_r M),\lambda\right)}{\prod_{i=1}^{r}\left(1 - \lambda^{\deg \theta_i}\right)}$$

5.7 DEFINITION. (i) If m = 1, let depth M := length of longest homogeneous M-sequence.

(ii) If m > 1, specialize the grading to a $\mathbb{Z}$-grading in any way and define depth M as in (i). (It can be shown that this definition is independent of the specialization.)

It is clear that depth M ≤ dim M. The case of equality, i.e., when some hsop is regular, is of particular importance.

5.8 DEFINITION. M is Cohen-Macaulay if depth M = dim M.

5.9 THEOREM. Let M have an hsop. Then M is C-M

⟺ every hsop is regular

⟺ M is a finitely-generated and free $k[\theta]$-module for some (equivalently, every) hsop $\theta = (\theta_1,\theta_2,\ldots,\theta_d)$.

5.10 THEOREM. Let M be C-M, with an hsop $\theta = (\theta_1,\theta_2,\ldots,\theta_d)$. Let $\eta_1,\eta_2,\ldots,\eta_t \in H(M)$. Then

$M = \coprod_{i=1}^{t} \eta_i k[\theta] \iff \eta_1, \eta_2, \ldots, \eta_t$ is a k-basis for $M/\theta M$. For such a choice of θ's and η's it follows that

$$F(M,\lambda) = \frac{\sum_{i=1}^{t} \lambda^{\deg \eta_i}}{\prod_{j=1}^{d} \left(1-\lambda^{\deg \theta_j}\right)}$$

Returning to our ring R_Φ once more, we can now state the following theorem, which will be proved later.

5.11 THEOREM (Hochster [H_1]). R_Φ is Cohen-Macaulay.

5.12 COROLLARY. $\sum_{r \geq 0} H_n(r)\lambda^r = \frac{P(\lambda)}{(1-\lambda)^{(n-1)^2+1}}$, $P(\lambda) \in \mathbb{N}[\lambda]$.

The corollary follows since permutation matrices have degree one. It is an open problem to compute $P(\lambda)$ or even $P(1)$ in a simple way. In particular, can $P(1)$ be computed more quickly than $P(\lambda)$?

5.13 THEOREM. Let $\dim R_\Phi = d$. There exist free commutative monoids $G_1, G_2, \ldots, G_t \subseteq E_\Phi$, all of rank d, and also $\eta_1, \eta_2, \ldots, \eta_t \in E_\Phi$, such that $E_\Phi = \biguplus_{i=1}^{t} (\eta_i + G_i)$, where $\uplus$ denotes disjoint union.

In terms of the ring this theorem says that $R_\Phi = \coprod_{i=1}^{t} x^{\eta_i} k[G_i]$. This is analogous to the C-M property, but differs in that the G_i's change. The proof is combinatorial, and uses the shellability of convex polytopes (due to Bruggesser and Mani). The proof is sketched in [St_{11}, §5].

5.14 EXAMPLE. For the equation $x_1 + x_2 - x_3 - x_4 = 0$ we get

$$R_\Phi = k[x_1x_3, x_1x_4, x_2x_3] \oplus x_2x_4 k[x_1x_4, x_2x_3, x_2x_4] .$$

Here x_1x_3 corresponds to the solution (1,0,1,0) as usual, and the geometry of the cone of solutions after triangulation is

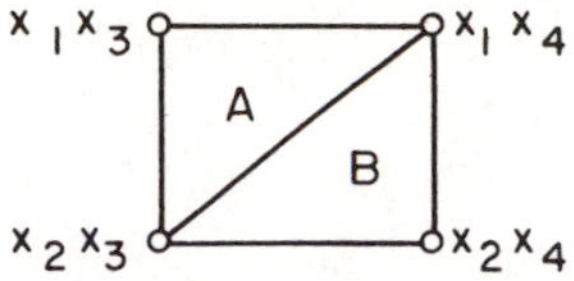

Geometrically, we have taken all integer points in cone A, and "pushed off" cone B from its intersection with A by translation by (0,1,0,1).

§6. Local cohomology

We now turn to the proof that R_Φ is Cohen-Macaulay, and more generally to the question of deciding depth $M_{\Phi,\alpha}$. For this we shall use the tool of local cohomology, which will first be reviewed. As always, all rings R and modules M are graded. Local cohomology will be defined with respect to the irrelevant ideal R_+.

Let $L_{R_+}(M) = L(M) = \{u \in M | R_+^n u = 0 \text{ for some } n > 0\}$. If $f : M \to N$, then $L(f) : L(M) \to L(N)$ by restriction. It is easy to check that L is a left-exact additive functor, so we can take the right-derived functors R^iL.

6.1 DEFINITION. $H^i(M) = H^i_{R_+}(M) = R^iL_{R_+}(M)$.

Some of the fundamental properties of local cohomology $H^i(M)$ will now be stated. In particular, the reader unfamiliar or unenamored with homological algebra can adopt the following theorem as the definition of $H^i(M)$. In the following R_{y_i} denotes R localized at y_i (i.e., with respect to the multiplicative set generated by y_i), and $M_{y_i} = M \otimes_R R_{y_i}$.

6.2 THEOREM. $H^i(M) = H^i\left[\bigotimes_{i=1}^{s} (0 \to R \to R_{y_i} \to 0) \otimes M\right]$

$$= H^i\left[0 \xrightarrow{\delta_0} M \xrightarrow{\delta_1} \coprod_i M_{y_i} \xrightarrow{\delta_2} \coprod_{i<j} M_{y_iy_j} \xrightarrow{\delta_3} \cdots \xrightarrow{\delta_s} M_{y_1\cdots y_s} \to 0\right]$$

$= \ker \delta_{i+1}/\mathrm{im}\ \delta_i$,

where $y_1, y_2, \ldots, y_s \in H(R_+)$ and $\mathrm{Rad}(y_1, \ldots, y_s) = R_+$.

The complex $\bigotimes_{i=1}^{s} (0 \to R \to R_{y_i} \to 0) \otimes M$ is denoted $K(\underline{y}^\infty, M)$. Local cohomology is depth sensitive in the following sense.

6.3 THEOREM. $H^i(M) = 0$ unless $e = \text{depth } M \le i \le \dim M = d$; and $H^e(M) \neq 0$, $H^d(M) \neq 0$.

Theorem 6.2 imposes on $H^i(M)$ in a natural way the structure of a $\mathbb{Z}^m$-graded module: $H^i(M) = \coprod_{\alpha \in \mathbb{Z}^m} H^i(M)_\alpha$. Since $H^i(M)$ is known to be artinian, it follows that $H^i(M)_\alpha = 0$ for $\alpha >> 0$. However, $H^i(M)$ is usually not finitely-generated. Define

$$F(H^i(M),\lambda) = \sum_{\alpha \in \mathbb{Z}^m} (\dim_k H^i(M)_\alpha)\lambda^\alpha .$$

6.4 THEOREM. $F(M,\lambda)_\infty = \sum_{i=e}^{d} (-1)^i F(H^i(M),\lambda)$.

In this formula $F(M,\lambda)_\infty$ signifies that $F(M,\lambda)$ is to be expanded as a Laurent series around ∞. For instance, for $R = k[x]$, $H^1(R) = x^{-1}k[x^{-1}]$ and $F(R,\lambda) = \frac{1}{1-\lambda} = -\frac{\lambda^{-1}}{1-\lambda^{-1}} = -\sum_{n<0} \lambda^n = -F(H^1(R),\lambda)$. Let $\deg F(R,\lambda)$ denote the degree as a rational function, i.e., degree of numerator minus degree of denominator.

6.5 COROLLARY. If $m = 1$ and $H^i(R)_\alpha = 0$ for all $\alpha > 0$ and all i, then $\deg F(R,\lambda) < 0$.

Proof. Write (uniquely) $F(R,\lambda) = G(R,\lambda) + L(R,\lambda)$ where $\deg G(R,\lambda) < 0$ and $L(R,\lambda) \in \mathbb{Z}[\lambda]$. The expansion $G(R,\lambda)_\infty$ has only negative exponents, while $L(R,\lambda)_\infty = L(R,\lambda)$. Hence if $H^i(R)_\alpha = 0$ for $\alpha > 0$ and all i, then by the previous theorem $F(R,\lambda) = G(R,\lambda)$. □

The condition $\deg F(R,\lambda) < 0$ is equivalent to saying that the Hilbert function $H(R,n)$ has no "exceptional" values. E.g., if R is generated by R_1 then $H(R,n)$ is a polynomial for all $n \ge 0$.

§7. Local cohomology of the modules $M_{\Phi,\alpha}$.

Let Φ be an $r \times n$ $\mathbb{Z}$-matrix and $\alpha \in \mathbb{Z}^n$ as before, and recall the definitions of E_Φ, $E_{\Phi,\alpha}$, R_Φ and $M_{\Phi,\alpha}$. Let $\overline{E}_\Phi$ denote the group generated by E_Φ in $\mathbb{Z}^n$, and let $\overline{E}_{\Phi,\alpha} := \overline{E}_\Phi + E_{\Phi,\alpha}$, the coset of $\overline{E}_\Phi$ in $\mathbb{Z}^n$ containing $E_{\Phi,\alpha}$. The nonnegative real solutions to $\Phi\beta = 0$ form a convex polyhedral cone $C_\Phi = \{\beta \in (\mathbb{R}^+)^n | \Phi\beta = 0\}$. Any cross-section of C_Φ, i.e., bounded intersection with a hyperplane meeting the relative interior of C_Φ, is a convex polytope, and different cross-sections are combinatorially equivalent. If F is a face of the cross-section polytope P_Φ, define supp F := supp β for any $\beta \in F - \partial F$, where as before $\text{supp}(\beta_1,\ldots,\beta_n) = \{i | \beta_i > 0\}$.

We now try to compute the local cohomology of R_Φ by considering the complex $K(\underline{y}^\infty, M)$, $M = M_{\Phi,\alpha}$. Recall the set of completely fundamental solutions $CF_\Phi = \{\delta_1,\delta_2,\ldots,\delta_s\}$, consisting of the integer points nearest the origin on the extreme rays of C_Φ. Thus, CF_Φ is in one-to-one correspondence with the vertices of the polytope P_Φ. Set $y_i = x^{\delta_i} \in R_\Phi$. We know by Theorem 3.7 that $\text{Rad}(y_1,\ldots,y_s) = R_+$. For $\beta \in \overline{E}_{\Phi,\alpha}$ consider the part of $K(\underline{y}^\infty, M)$ of degree β:

$$0 \to M_\beta \to \coprod_i (M_{y_i})_\beta \to \coprod_{i<j} (M_{y_i y_j})_\beta \to \cdots \quad . \tag{1}$$

In, say, $M_{y_i y_j}$ we have inverted everything on the face spanned by y_i and y_j. So, to get something of degree β in $M_{y_i y_j}$ take anything in $E_{\Phi,\alpha}$ and subtract δ_i and δ_j any number of times (i.e., multiply by $x^{-\delta_i}$ and $x^{-\delta_j}$). In general, there is something in $M_{y_{i_1}\cdots y_{i_r}}$ of degree β if and only if $N(\delta_{i_1} + \cdots + \delta_{i_r}) + \beta \geq 0$ for $N \gg 0$. This condition is equivalent to $\text{supp}_-\, \beta \subseteq \text{supp}\, F$, where F is the face spanned by $\delta_{i_1},\ldots,\delta_{i_r}$ and $\text{supp}_-(\beta_1,\ldots,\beta_n) = \{i | \beta_i < 0\}$, the <u>negative support</u> of β. Thus

$$\dim_k (M_{y_{i_1}\cdots y_{i_r}})_\beta = \begin{cases} 1\ , & \text{if } \text{supp}_-\, \beta \subseteq \text{supp}\, F \\ 0\ , & \text{otherwise} \quad . \end{cases}$$

So all the pieces of the β-part (1) of $K(\underline{y}^\infty, M)$ are 0- or 1-dimensional vector spaces. Now, the key fact is that we can identify the complex (1) with the augmented chain complex of the simplicial complex Δ_β whose faces are the sets $S \subseteq CF_\Phi$ such that $\text{supp}_- \beta \subseteq \bigcup_{\delta \in CF_\Phi - S} (\text{supp}\ \delta) = \text{supp}$ (face F of P_Φ spanned by all $\delta \in CF_\Phi - S$). We may therefore compute local cohomology by computing reduced simplicial homology. For further details, see [St_{11}].

7.1 EXAMPLE. Consider the equation $x_1 + x_2 - x_3 - x_4 = 2$, i.e., let $\Phi = [1\ 1\ -1\ -1]$ and $\alpha = 2$. Here $\text{FUND}_\Phi = CF_\Phi$ consists of the four elements (1,0,1,0), (1,0,0,1), (0,1,0,1), (0,1,1,0), so $y_1 = x_1x_3$, $y_2 = x_1x_4$, $y_3 = x_2x_4$ and $y_4 = x_2x_3$. Geometrically, P_Φ is a square. Let $\beta = (0,0,-1,-1)$ and $M = M_{\Phi,\alpha}$, and write $M_i = (M_{y_i})_\beta$, $M_{ij} = (M_{y_iy_j})_\beta$, and so on. Then the complex (1) takes the following form, where the non-zero pieces are underlined:

$$0 \to M \to M_1 \oplus M_2 \oplus M_3 \oplus M_4 \to$$

$$\to \underline{M_{12}} \oplus \underline{M_{13}} \oplus M_{14} \oplus M_{23} \oplus \underline{M_{24}} \oplus \underline{M_{34}} \to$$

$$\to \underline{M_{123}} \oplus \underline{M_{124}} \oplus \underline{M_{134}} \oplus \underline{M_{234}} \to \underline{M_{1234}} \to 0 .$$

Taking complements of the underlined elements we find that $\Delta_\beta = \{34,24,13,12,4,3,2,1,\emptyset\}$, and the homology of Δ_β (a circle) is non-zero only in dimension one, corresponding to the $M_{12} \oplus \cdots$ part of (1). In terms of local cohomology this means that $\dim_k(H^2(M_{\Phi,\alpha})_\beta) = 1$, so $M_{\Phi,\alpha}$ is not Cohen-Macaulay, since $\dim M_{\Phi,\alpha} = 3$.

Let P^* be the dual polytope of $P = P_\Phi$, and define $\Gamma_\beta = \cup F^*$, the union over all faces F of P such that $\text{supp}_-\beta \subseteq \text{supp}\ F$. Thus, Γ_β is a polyhedral subcomplex of P^*. Let $d = \dim R_\Phi$, so $\dim P = d - 1$, and let $s = |CF_\Phi|$. Our next result shows that we can replace the simplicial complex Δ_β by the more tractable Γ_β.

7.2 THEOREM. $\tilde{H}_i(\Gamma_\beta;k) \cong \tilde{H}_{s-d+i}(\Delta_\beta;k)$.

The proof uses Alexander duality twice and also a theorem on lattice homology of J. Folkman [F]. See [St_{11}, Lemma 2.8]. In view of this result we conjecture that Δ_β has the homotopy type of the (s-d)-fold suspension of Γ_β.

7.3 THEOREM. Let $M = M_{\Phi,\alpha}$, $d = \dim M$. Then as vector spaces over k,

$$H^i(M)_\beta \cong \begin{cases} \tilde{H}_{d-1-i}(\Gamma_\beta;k) , & \text{if } \beta \in \overline{E}_{\Phi,\alpha} \\ 0 , & \text{otherwise} . \end{cases}$$

Keeping in mind that $\tilde{H}_1(\phi;k) = k$ if $i = -1$ and $= 0$ otherwise, we deduce the following.

7.4 COROLLARY. M is Cohen-Macaulay $\Longleftrightarrow$ for all $\beta \in \overline{E}_{\Phi,\alpha}$ either $\Gamma_\beta = \emptyset$ or Γ_β is acyclic.

7.5 COROLLARY. $H^d(M)_\beta \cong \begin{cases} k , & \text{if } \beta \in \overline{E}_{\Phi,\alpha} \text{ and } \Gamma_\beta = \emptyset \\ 0 , & \text{otherwise} . \end{cases}$

For the C-M condition above note that $\Gamma_\beta = \emptyset$ if and only if $\mathrm{supp}_-\ \beta \subseteq \mathrm{supp}\ F$ implies $F = P$. Also, if $\alpha = 0$ then $\Gamma_\beta = \emptyset \Longleftrightarrow \mathrm{supp}_-\beta = \mathrm{supp}\ P$ (there is a simple geometric proof for this). In particular, if there exists $\gamma \in E_\Phi$ such that $\gamma > 0$, then for $\beta \in \overline{E}_\Phi$: $\Gamma_\beta = \emptyset \Longleftrightarrow \beta < 0$ (strict inequality in all coordinates). Hence, for $\alpha = 0$ there is the following "reciprocity" result (when $E_\Phi \cap \mathbb{P}^n \neq \emptyset$):

R_Φ is spanned by <u>nonnegative</u> solutions to $\Phi\beta = 0$, and

$H^d(R_\Phi)$ is spanned by <u>strictly negative</u> solutions to $\Phi\beta = 0$.

In general, $H^d(M_{\Phi,\alpha}) \cong k\{x^\beta | \beta \in \overline{E}_{\Phi,\alpha} , \Gamma_\beta = \emptyset\}$, with the R_Φ-module structure given by

$$x^\gamma \cdot x^\beta = \begin{cases} x^{\gamma+\beta} , & \text{if } \Gamma_{\gamma+\beta} = \emptyset . \\ 0 , & \text{otherwise} , \end{cases}$$

for $\gamma \in E_\Phi$, $x^\beta \in H^d(M_{\Phi,\alpha})$.

7.6 COROLLARY. (a) (Hochster [H_1]) R_Φ is Cohen-Macaulay, and (b) for any specialization of R_Φ to an $\mathbb{N}$-grading, $\deg F(R_\Phi,\gamma) < 0$.

Proof. (a) We must show that if $\beta \in \overline{E}_\Phi$ and $\Gamma_\beta \neq \emptyset$ then Γ_β is acyclic. Assume that β satisfies these hypotheses. We will sketch an argument that Γ_β is in fact a ball. We may assume $\beta \notin E_\Phi$, since if $\beta \in E_\Phi$ then $\Gamma_\beta = P^*$. Stand at point β in the d-dimensional vector space spanned by the cone C_Φ, and look at P. It is not hard to show, since $\Gamma_\beta \neq \emptyset$, that the faces of P visible from β (regarding C_Φ as opaque) form a (d-2)-ball. The faces F you don't see are those such that $\text{supp}_-\,\beta \subseteq \text{supp}\,F$. Passing to the dual polytope P^*, the non-visible part of P goes to Γ_β. By polytopal duality Γ_β is a ball (and hence acyclic).

(b) Part (a) gives that $F(R_\Phi,\lambda)_\infty = (-1)^d F(H^d(R_\Phi),\lambda)$. Thus, $F(R_\Phi,\lambda)_\infty$ has no exponents ≥ 0, so as before we conclude that the Hilbert function of R_Φ has no exceptional values. □

Thus we have also proved Corollary 5.12. An analogue of Corollary 5.12 for symmetric magic squares follows in the same way.

While the modules $M_{\Phi,\alpha}$ may fail to be Cohen-Macaulay in general, some of them which are "close enough" to R_Φ are in fact C-M. Let us mention some results in this direction.

7.7 COROLLARY. Suppose there exists $\gamma = (\gamma_1,\ldots,\gamma_n) \in \mathbb{Q}^n$, $-1 < \gamma_i \leq 0$, such that $\Phi\gamma = \alpha$. Then $M_{\Phi,\alpha}$ is Cohen-Macaulay.

Proof. Let $\Phi\beta = \alpha$, $\beta \in \overline{E}_{\Phi,\alpha}$. Then $\Phi(\beta-\gamma) = 0$ and $\text{supp}_-\beta = \text{supp}_-(\beta-\gamma)$. Γ_β depends only on $\text{supp}_-\beta$, so $\Gamma_\beta = \Gamma_{m(\beta-\gamma)}$, where m is chosen so that $m(\beta-\gamma)$ is integral. Since R_Φ is C-M the complex $\Gamma_{m(\beta-\gamma)}$ is empty or acyclic. Hence, so is Γ_β and we are done. □

As an aside we mention that given $\Phi = [\rho_1\rho_2\cdots\rho_n]$ the number of α's satisfying the hypotheses of the previous corollary equals

$$\sum_{\substack{\rho_{i_1}\cdots\rho_{i_j}\\ \text{lin.indep.}}} (-1)^{r-j}\ \text{g.c.d.}\ (j \times j\text{-minors of } [\rho_{i_1}\cdots\rho_{i_j}]) \ .$$

7.8 COROLLARY. Let $\Phi = [a_1,\ldots,a_s,-b_1,\ldots,-b_t]$, $a_i,b_j > 0$, $s,t > 0$. Thus, $d = \dim R_\Phi = s + t - 1$. Let $\Phi[{x \atop y}] = \alpha$ stand for $\Sigma\, a_i x_i - \Sigma\, b_j y_j = \alpha$. Then for $0 \leq i < d$:

$$H^i(M_{\Phi,\alpha}) \cong \begin{cases} k, \text{ if } i = s,\ \Phi\left[\begin{smallmatrix} x \\ y \end{smallmatrix}\right] = \alpha,\ x < 0, \text{ and } y \geq 0, \\ k, \text{ if } i = t,\ \Phi\left[\begin{smallmatrix} x \\ -y \end{smallmatrix}\right] = \alpha,\ x \geq 0, \text{ and } y < 0, \\ 0, \text{ otherwise.} \end{cases}$$

Consequently,

$$\operatorname{depth} M_{\Phi,\alpha} = \begin{cases} 0 & , \text{ if } M_{\Phi,\alpha} = 0 \\ s & , \text{ if } \exists \text{ solution } x < 0,\ y \geq 0 \\ t & , \text{ if } \exists \text{ solution } x \geq 0,\ y < 0 \\ s + t - 1 & , \text{ otherwise.} \end{cases}$$

The proof is based on topology - it is possible to compute the complexes Γ_β explicitly. Note that the middle two conditions are mutually exclusive; if there exists a solution $x < 0$, $y \geq 0$ then $\alpha < 0$ and in the other case $\alpha > 0$.

In conclusion we mention the following result, though no applications of it have yet been found. It was discovered independently by M. Hochster, the author, and perhaps others.

7.9 THEOREM. Let $E \subseteq \mathbb{N}^m$ be a finitely-generated monoid. Let $G \subseteq \mathbb{Z}^m$ be a finitely-generated "E-module", i.e., $E + G \subseteq G$. Let $R = kE$ and $M = kG$. Let $V = \{\beta_1, \beta_2, \ldots, \beta_t\} \subseteq E$ such that R is integral over $k[x^{\beta_1}, x^{\beta_2}, \ldots, x^{\beta_t}]$ (the subalgebra generated by the x^{β_i}'s). Let $A = k[y_1, y_2, \ldots, y_t] \to k[x^{\beta_1}, x^{\beta_2}, \ldots, x^{\beta_t}]$ be the surjection given by $y_i \mapsto x^{\beta_i}$. So M is a finitely-generated A-module. If $\gamma \in G$, define a simplicial complex Δ_γ on the vertex set V having faces $\{\beta_{i_1}, \ldots, \beta_{i_r}\}$ such that $\gamma - \beta_{i_1} - \cdots - \beta_{i_r} \in G$. Then the Betti numbers $\beta_i^A(M) := \dim_k \operatorname{Tor}_i^A(M,k)$ satisfy

$$\beta_i^A(M) = \sum_{\gamma \in G} \dim_k \tilde{H}_{i-1}(\Delta_\gamma; k) .$$

§8. Reciprocity

Some reciprocity theorems in the theory of linear diophantine equations which were originally proved by combinatorial methods find a pleasing explanation in the setting of Cohen-Macaulay modules and local cohomology. For instance, the formula

$$F(R_\Phi,\lambda)_\infty = (-1)^d \sum_{\substack{\beta\in\overline{E}_\Phi \\ \text{supp}\_\beta\, =\, \text{supp}\, P}} \lambda^\beta$$

first proved in 1973 [St_1, Thm. 4.1] now follows from the reciprocity of local cohomology (Theorem 6.4), the Cohen-Macaulayness of R_Φ (Corollary 7.6), and Corollary 7.5. In the same way we could derive the following formula, implicit in [St_2, Thm. 10.2]: If there exists a $\gamma = (\gamma_1,\ldots,\gamma_n) \in \mathbb{Q}^n$, $-1 < \gamma_i \leq 0$, such that $\Phi\gamma = \alpha$, then

$$F(M_{\Phi,\alpha}, \lambda)_\infty = (-1)^d \sum_{\substack{\beta\in\overline{E}_{\Phi,\alpha} \\ \text{supp}\_\beta\, =\, \text{supp}\, P}} \lambda^\beta \,. \qquad (2)$$

In fact, it is clear from the previous section that the following more general statement is valid.

8.1 THEOREM. If $M_{\Phi,\alpha}$ is Cohen-Macaulay then (2) holds.

Leaving the C-M case the formula (2) will in general fail. However, using the local cohomology expansion one can still get an exact formula revealing the error.

8.2 RECIPROCITY THEOREM. Let $d = \dim M_{\Phi,\alpha}$ and $e = \operatorname{depth} M_{\Phi,\alpha}$. Then

$$F(M_{\Phi,\alpha},\lambda)_\infty = \sum_{i=e}^{d} (-1)^i \left[\sum_{\beta\in\overline{E}_{\Phi,\alpha}} \left(\dim_k \tilde{H}_{d-i-1}(\Gamma_\beta;k)\right)\lambda^\beta \right]$$

$$= \sum_{\beta\in\overline{E}_{\Phi,\alpha}} \tilde{\chi}(\Gamma_\beta)\lambda^\beta \,,$$

where $\tilde{\chi}$ denotes reduced Euler characteristic.

Proof. Insert the formula for $F(H^i(M_{\Phi,\alpha}),\lambda)$ obtained from Theorem 7.3 into the reciprocity formula of local cohomology (Theorem 6.4). □

The main term $(i = d)$ corresponds to the right hand side of (2) and the other terms are corrections. It is now evident when these corrections vanish. In particular, C-M-ness is not necessary for (2).

8.3 COROLLARY. (2) holds $\Longleftrightarrow$ for all $\beta \in \overline{E}_{\Phi,\alpha}$ such that $\Gamma_\beta \neq \emptyset$, $\tilde{\chi}(\Gamma_\beta) = 0$.

Let $P(n) \in \mathbb{C}[n]$, $\deg P = d-1$, and let $F(x) = \sum_{n\geq 0} P(n)x^n = W(x)(1-x)^{-d}$, $\deg W < d$. The following reciprocity theorem is actually true not just for polynomials $P(n)$, but for functions $\Sigma_{i=1}^{s} P_i(n)\gamma_i^n$, where $P_i(n) \in \mathbb{C}[n]$ and $0 \neq \gamma_i \in \mathbb{C}$. However, we will only need the polynomial case.

8.4 THEOREM (Popoviciu [Po, p. 8]). $F(x)_\infty = -\sum_{n<0} P(n)x^n$.

Proof. Consider the $\mathbb{C}[x]$-module $\{\sum_{n\in\mathbb{Z}} f(n)x^n \mid f : \mathbb{Z} \to \mathbb{C}\}$. Let $G(x) = \sum_{n\in\mathbb{Z}} P(n)x^n$. Since the d-th difference is 0 for a polynomial of degree $d - 1$, i.e., $\sum_{i=0}^{d} (-1)^{d-i}\binom{d}{i}P(n+i) = 0$ for all n, we get $(1-x)^d G(x) = 0$. Hence, $W(x) = (1-x)^d F(x) = -(1-x)^d \sum_{n<0} P(n)x^n$ which shows that $F(x)$ and $-\sum_{n<0} P(n)x^n$ are equal as rational functions. □

§9. Reciprocity for integer stochastic matrices

Let us once more return to the problem of enumerating magic squares. Recall that $H_n(r)$ denotes the polynomial which for $r \geq 0$ counts the number of $n \times n$ $\mathbb{N}$-matrices having line sum r for all lines (i.e., rows and columns). Also, for $r < 0$ let $\overline{H}_n(r)$ be the number of $n \times n$ matrices of strictly negative integers having constant line sum r. Set $F(x) := \sum_{r\geq 0} H_n(r)x^r$. By our earlier reciprocity results

$$F(x)_\infty = (-1)^{(n-1)^2+1} \sum_{r<0} \bar{H}_n(r)x^r ,$$

and by the theorem of Popoviciu (Theorem 8.4)

$$F(x)_\infty = - \sum_{r<0} H_n(r)x^r .$$

Hence, $\bar{H}_n(r) = (-1)^{n-1}H_n(-r)$ for all $r < 0$. There exists a simple transformation between the positive and strictly negative cases as follows: M is an $\mathbb{N}$-matrix with line sum $r \Longleftrightarrow -M -J$ is a matrix of strictly negative integers having line sum $-n-r$, where J is the $n\times n$ matrix of all one's. This bijection shows that $H_n(r) = \bar{H}_n(-n-r)$. Hence, we conclude that

$$H_n(r) = (-1)^{n-1}H_n(-n-r), \text{ and}$$

$$H_n(-1) = H_n(-2) = \cdots = H_n(-n+1) = 0.$$

All parts (i) - (iv) of the Anand-Dumir-Gupta conjecture have now been verified. We remark that with this information it is possible to explicitly determine the polynomials $H_n(r)$ for small values of n. For instance, for the case $n = 3$ (first done by MacMahon) we know that $H_3(-1) = H_3(-2) = 0$, $H_3(0) = H_3(-3) = 1$ and $H_3(1) = H_3(-4) = 6$, and being a polynomial of degree $(3-1)^2 = 4$ these six values determine $H_3(r)$. With the aid of a computer $H_n(r)$ has been explicitly computed up to $n = 6$ [J-V].

§10. Rational points in integral polytopes

A topic closely related to the theory of linear diophantine equations is that of integral convex polytopes, which we will now mention in passing. Let $P \subseteq \mathbb{R}^n$ be a convex d-dimensional polytope with vertices in $\mathbb{Z}^n$. For $m > 0$ let

$$i(P,m) = \#\{\alpha \in P \mid m\alpha \in \mathbb{Z}^n\} \quad ,$$

$$\bar{i}(P,m) = \#\{\alpha \in P - \partial P \mid m\alpha \in \mathbb{Z}^n\} \quad .$$

For instance, if P is the square in $\mathbb{R}^2$ having vertices (0,0), (1,0), (0,1) and (1,1) then $i(P,m) = (m+1)^2$ and $\bar{i}(P,m) = (m-1)^2$. The following result is due to Ehrhart [E_1] [E_2] (made more precise by Macdonald [Md_1] [Md_2]). It also follows from [St_1], was proved independently in [Mc_2] and [St_9], and is a simple consequence of Corollary 3.7, Corollary 7.6(b), Theorem 8.1, and Theorem 8.4.

10.1 THEOREM. $i(P,m)$ and $\bar{i}(P,m)$ are polynomials of degree d, $i(P,0) = 1$, and $\bar{i}(P,m) = (-1)^d i(P,-m)$.

10.2 EXAMPLE. Let $P = \Omega_n$ = {doubly stochastic $n \times n$ matrices} $\subseteq \mathbb{R}^{n^2}$. Ω_n is a convex polytope of dimension $(n-1)^2$, and by the Birkhoff-Von Neumann theorem its vertices are the permutation matrices. One finds that $i(\Omega_n,m) = \#\{n \times n$ $\mathbb{N}$-matrices with constant line sum $m\} = H_n(m)$ and $\bar{i}(\Omega_n,m) = \#\{n \times n$ $\mathbb{P}$-matrices with constant line sum $m\} = \bar{H}_n(-m)$. Thus the earlier results on the enumeration of magic squares (Anand-Dumir-Gupta conjecture) can also be derived via the preceding theorem.

10.3 THEOREM. Let $d = n$. Then $i(P,m) = (\text{vol } P)m^d$ + lower terms.

<u>Proof</u>. Fix $m > 0$. For every $\alpha \in P$ such that $m\alpha \in \mathbb{Z}^d$, surround α by a d-dimensional cube of side m^{-1}. There are by definition $i(P,m)$ such cubes, each of volume m^{-d}. Hence, essentially by the definition of the Riemann integral $\lim_{m\to\infty} i(P,m)m^{-d} = \text{vol } P$. □

10.4 COROLLARY. Any d values of $i(P,m)$, $m \geq 1$, or $\bar{i}(P,m)$, $m \geq 1$, determine the volume of P.

For instance, for $d = 2$ vol P is determined by $i(P,1)$ and $\bar{i}(P,1)$. Such a result was proved by Pick [Pi] for nonconvex polygonal regions, and Pick's theorem can be obtained by subdivision from the convex case considered here. Observe that in terms of $d + 1$ consecutive values of

$i(P,m)$ we have

$$\text{vol } P = \frac{1}{d!} \sum_{j=0}^{d} (-1)^{d-j} \binom{d}{j} i(P, m + j) .$$

§11. Free resolutions

For further developments we shall need to review again some algebraic background. Let R be a finitely-generated $\mathbb{N}^m$-graded k-algebra and M a finitely-generated $\mathbb{Z}^m$-graded R-module. As usual M can be regarded as a finitely-generated module over a polynomial ring A. If $x_1, x_2, \ldots, x_s \in H(R_+)$ generate R make the surjection $A = k[y_1, y_2, \ldots, y_s] \to R$, $y_i \mapsto x_i$, degree-preserving by setting $\deg y_i = \deg x_i$.

11.1 DEFINITION. A finite free resolution (f.f.r.) of M (as a graded A-module) is an exact sequence

$$0 \to \Lambda_t \xrightarrow{\phi_t} \Lambda_{t-1} \xrightarrow{\phi_{t-1}} \cdots \xrightarrow{\phi_2} \Lambda_1 \xrightarrow{\phi_1} \Lambda_0 \xrightarrow{\phi_0} M \to 0 \qquad (3)$$

where the Λ_i's are free finitely-generated graded A-modules:

$$\Lambda_i \cong A(\alpha_{1i}) \oplus A(\alpha_{2i}) \oplus \cdots \oplus A(\alpha_{q_i i}) , \quad \alpha_{ji} \in \mathbb{Z}^m , \text{ and}$$

$$A(\alpha) \cong A \text{ with grading } A(\alpha)_\beta \cong A_{\beta-\alpha} ,$$

and where the maps ϕ_i are degree-preserving.

The homological dimension of M, $\text{hd}_A M$, is the minimal t possible in (3). By the Hilbert syzygy theorem $\text{hd}_A M \leq s = \dim A$. The following sharper result is due to Auslander and Buchsbaum.

11.2 THEOREM. $\text{hd}_A M = s - \text{depth } M$.

The f.f.r. (3) is said to be minimal if each Λ_i has smallest possible rank (it can easily be shown that these ranks can be simultaneously minimized). If (3) is minimal let $\beta_i^A(M) := \text{rank } \Lambda_i$. An equivalent definition of the Betti numbers β_i is $\beta_i^A(M) = \dim_k \text{Tor}_i^A(M,k)$.

The Hilbert series can be read off immediately from any finite free resolution (3).

11.3 THEOREM. $F(M,\lambda) = \sum_{i=0}^{t} (-1)^i F(\Lambda_i,\lambda)$

$$= \frac{\sum_{i=0}^{t} (-1)^i (\lambda^{\alpha_{1i}} + \cdots + \lambda^{\alpha_{q_i i}})}{\prod_{j=1}^{s} (1 - \lambda^{\deg y_j})} .$$

11.4 **EXAMPLE.** Let $A = k[x,y,z,w]$, $R = A/(xyz,xw,yw,zw)$, and for simplicity let $\deg x = \deg y = \deg z = \deg w = 1$. Consider the following f.f.r. of R:

$$0 \to A \xrightarrow{\phi_3} A^4 \xrightarrow{\phi_2} A^4 \xrightarrow{\phi_1} A \to R \to 0$$

$$\phi_3 = [z \ \ -y \ \ x \ \ 0] \qquad \phi_2 = \begin{bmatrix} 0 & y & -x & 0 \\ 0 & z & 0 & -x \\ 0 & 0 & z & -y \\ w & -yz & 0 & 0 \end{bmatrix} \qquad \phi_1 = \begin{bmatrix} xyz \\ xw \\ yw \\ zw \end{bmatrix}$$

This f.f.r. can quickly be seen to be minimal (assuming it is indeed an f.f.r.) using the following criterion.

11.5 PROPOSITION. An f.f.r. is minimal $\Leftrightarrow$ no matrix entry $\in k^*$.

The degrees of basis elements in the above f.f.r. are

$$0 \longrightarrow \underset{4}{A} \longrightarrow \underset{3,3,3,4}{A^4} \longrightarrow \underset{3,2,2,2}{A^4} \longrightarrow \underset{0}{A} \longrightarrow R \longrightarrow 0$$

so we can read off the Hilbert series

$$F(R,\lambda) = \frac{1 - (3\lambda^2 + \lambda^3) + (3\lambda^3 + \lambda^4) - \lambda^4}{(1 - \lambda)^4}$$

Now $hd_A R = 3 = 4 - \text{depth } R$, so depth R = 1. But dim R = 2, so R is not Cohen-Macaulay.

If a Hilbert series has been computed by other means it is not in general possible to decompose as to see the Betti numbers. For instance, with A and R as in the previous example let $R_1 = A/(xz,xw,yw)$. Then $F(R_1,\lambda) = F(R,\lambda) = (1-\lambda)^{-4}(1 - 3\lambda^2 + 2\lambda^3)$, but R_1 is Cohen-Macaulay and R is not. From the numerator $1 - 3\lambda^2 + 2\lambda^3$ of $F(R,\lambda)$ it is impossible to determine the "correct" decomposition $1 - (3\lambda^2 + \lambda^3) + (3\lambda^3 + \lambda^4) - \lambda^4$.

§12. Duality and canonical modules

An f.f.r. (3) can be dualized by applying the functor $\mathrm{Hom}_A(\cdot,A)$. If $\Lambda = A(\alpha_1) \oplus \cdots \oplus A(\alpha_q)$ then $\Lambda^* = \mathrm{Hom}_A(\Lambda,A) = A(-\alpha_1) \oplus \cdots \oplus A(-\alpha_q)$, so Λ^* is a free module of the same rank but with a different grading. All arrows are reversed, and the matrices expressing these arrows change to their transposes. For instance, the dual of the f.f.r. of the preceding example is as follows:

$$0 \xrightarrow{\phi_0^*} A \xrightarrow[\begin{bmatrix} xyz & xw & yw & zw \end{bmatrix}]{\phi_1^*} A^4 \xrightarrow[\begin{bmatrix} 0 & 0 & 0 & w \\ y & z & 0 & -yz \\ -x & 0 & z & 0 \\ 0 & -x & -y & 0 \end{bmatrix}]{\phi_2^*} A^4 \xrightarrow[\begin{bmatrix} z \\ -y \\ x \\ 0 \end{bmatrix}]{\phi_3^*} A \xrightarrow{\phi_4^*} 0$$

Clearly $\phi_{i+1}^*\phi_i^* = 0$, so the dualized resolution is a complex, but it is in general not exact. The homology of a dualized (minimal) free resolution of an R-module M, considered as A-module, is one of the fundamental functors of homological algebra (which is independent of the free resolution, minimal or not).

12.1 DEFINITION. $\mathrm{Ext}_A^i(M,A) = \ker \phi_{i+1}^* \,/\, \mathrm{im}\, \phi_i^*$.

The <u>injective</u> <u>hull</u> of k as an $A = k[y_1,\ldots,y_s]$-module is $E_A(k) = k[y_1^{-1},\ldots,y_s^{-1}]$. This given we define the (Matlis) <u>dual</u> <u>module</u> of any $\mathbb{Z}^m$-graded A-module M by

$$M^\vee = \mathrm{Hom}_A(M,E_A(k)) \ .$$

$M^\vee$ is made into a graded module by saying that $\phi : M \to E_A(k)$ has degree

α if $\phi(M_\beta) \subseteq E_A(k)_{\beta-\alpha}$ for all $\beta \in \mathbb{Z}^m$.

Fact 1. $F(M^\vee,\lambda) = F(M,\lambda^{-1})$.

Fact 2. $M^{\vee\vee} = \hat{M}$, the A_+-adic completion of M.

12.2 EXAMPLE. Let $A = k[x,y]$, $R = A/(xy)$, deg $x = (1,0)$ and deg $y = (0,1)$. A homogeneous k-basis for $R^\vee$ consists of those $\phi : R \to k[x^{-1},y^{-1}]$ such that $\phi(1) = 1$ or $\phi(1) = x^{-n}$, $n > 0$, or $\phi(1) = y^{-n}$, $n > 0$, since we cannot have negative exponents of both x and y appear in the image of the element 1. Thus,

$$F(R^\vee,\lambda) = 1 + \sum_{n>0} (\lambda_1^{-n} + \lambda_2^{-n}) \quad , \text{ and}$$

$$F(R,\lambda) = 1 + \sum_{n>0} (\lambda_1^{n} + \lambda_2^{n}) \quad .$$

The functors Ext and $^\vee$ are related to local cohomology by the following remarkable result [Ha, §6].

12.3 LOCAL DUALITY THEOREM. $\mathrm{Ext}_A^i(M,A)^\vee = H^{s-i}(M)$.

Let M be a Cohen-Macaulay module of dimension d with a minimal free resolution

$$0 \longrightarrow \Lambda_t \xrightarrow{\phi_t} \Lambda_{t-1} \xrightarrow{\phi_{t-1}} \cdots \xrightarrow{\phi_1} \Lambda_0 \longrightarrow M \longrightarrow 0.$$

Let $\Omega(M) = \mathrm{coker}\ \phi_t^* = \Lambda_t^* / \mathrm{im}\ \phi_t^* = \mathrm{Ext}_A^{s-d}(M,A)$. Equivalently, $\Omega(M)$ is the unique finitely-generated R-module whose completion $\hat{\Omega}(M) = \Omega(M) \otimes_R \hat{R}$ is isomorphic to $H^d(M)^\vee$. Then

$$0 \longrightarrow \Lambda_0^* \xrightarrow{\phi_1^*} \Lambda_1^* \longrightarrow \cdots \xrightarrow{\phi_t^*} \Lambda_t^* \longrightarrow \Omega(M) \to 0 \qquad (4)$$

is an exact sequence, because C-M-ness ensures that $H^i(M) \neq 0$ only for $i = d$, hence by local duality $\mathrm{Ext}_A^i(M,A) \neq 0$ only for $i = s - d = \mathrm{hd}_A M = t$. In fact, (4) is a minimal free resolution of $\Omega(M)$. $\Omega(M)$ is called the canonical module of M, and it can be shown directly that as an R-module $\Omega(M)$ is independent of A. It is seen from (4) that the Betti numbers of $\Omega(M)$ are the reverse of those of M:

$$\beta_i^A(\Omega(M)) = \beta_{h-i}^A(M) , \quad h = hd_A M .$$

$\Omega(M)$ has a natural $\mathbb{Z}^m$-grading such that $F(\Omega(M),\lambda) = (-1)^d F(M,\frac{1}{\lambda})$ as rational functions. In the following table we record the way that the Hilbert series varies with the fundamental modules associated with the Cohen-Macaulay module M. The subscripts 0 and ∞ signify expansion of a rational function around the origin and infinity respectively.

Module	Hilbert series
M	$F(M,\lambda)_0 = \sum_\alpha h_\alpha \lambda^\alpha$
$M^\vee$	$F(M,\frac{1}{\lambda})_\infty = \sum_\alpha h_\alpha \lambda^{-\alpha}$
$\Omega(M)$	$(-1)^d F(M,\frac{1}{\lambda})_0 = \sum_\alpha \bar{h}_\alpha \lambda^\alpha$
$\Omega(M)^\vee = H^d(M)$	$(-1)^d F(M,\lambda)_\infty = \sum_\alpha \bar{h}_\alpha \lambda^{-\alpha}$

Define the socle of a module M by soc $M := \{u \in M \mid R_+ u = 0\}$. It follows from noetherianness that $\dim_k$ soc $M < \infty$.

12.4 THEOREM. Let M be a Cohen-Macaulay module of dimension d over $A = k[x_1,\ldots,x_s]$. Then the following numbers are equal:

(a) $\beta_{s-d}^A(M)$

(b) the minimum number of generators of $\Omega(M)$ (as an A-module or an R-module)

(c) $\dim_k$ soc $H^d(M)$

(d) $\dim_k$ soc $M/(\theta_1 M + \cdots + \theta_d M)$ for any hsop $\theta_1,\ldots,\theta_d$.

Proof. The equivalence of (a), (b) and (c) follows from properties mentioned earlier, such as $\hat{\Omega}(M) = H^d(M)^\vee$, etc. (c) equals (d) by straightforward use of the long exact sequence for local cohomology. □

The number just characterized is called the type of M.

12.5 THEOREM. Let R = A/I be Cohen-Macaulay. Then the following are equivalent:

(a) type R = 1 ,

(b) $\Omega(R) \cong R$ (up to a shift in grading) .

A Cohen-Macaulay ring of type one is said to be Gorenstein. Thus a minimal free resolution of a Gorenstein ring is "self-dual". In particular,

$$\beta_i^A(R) = \beta_{s-d-i}^A(R) .$$

12.6 THEOREM. If R is Gorenstein then for some $\alpha \in \mathbb{Z}^m$,

$$F\left(R,\tfrac{1}{\lambda}\right) = (-1)^d \lambda^\alpha F(R,\lambda) .$$

Proof. $F\left(R,\tfrac{1}{\lambda}\right)_0 = (-1)^d F(\Omega(R),\lambda) = (-1)^d \lambda^\alpha F(R,\lambda)$. □

If m = 1 and R is Gorenstein with Hilbert series

$$F(R,\lambda) = \frac{h_0 + h_1\lambda + \cdots + h_t\lambda^t}{\prod_{i=1}^{d} (1 - \lambda^{\gamma_i})} , \quad h_t \neq 0 ,$$

then by the previous theorem $h_i = h_{t-i}$, $i = 0,1,\ldots,t$, and $\alpha = t - \Sigma\gamma_i = \deg F = \max\{j \mid H^d(R)_j \neq 0\}$. Also, if $\alpha \geq 0$ and each $\gamma_i = 1$ then α is the last value where the Hilbert function and the Hilbert polynomial disagree.

The converse to the preceding theorem is false. For instance, the ring $k[x,y]/(x^3,xy,y^2)$, deg x = deg y = 1, is Cohen-Macaulay, artinian and $F(R,\lambda) = \lambda^2 F\left(R,\tfrac{1}{\lambda}\right)$, but is not Gorenstein. For a reduced counterexample one can take $k[x,y,z,w]/(xyz,xw,yw)$. In the positive direction the following can be said [St_6, Thm. 4.4].

12.7 THEOREM. If R is a Cohen-Macaulay domain, then R is Gorenstein $\Longleftrightarrow F(R,\lambda) = (-1)^d \lambda^\alpha F(R,\frac{1}{\lambda})$ for some $\alpha \in \mathbb{Z}^m$.

Let us now review a few more facts about canonical modules of Cohen-Macaulay rings. The basic reference is [H-K].

12.8 THEOREM. $\Omega(R)$ is isomorphic to an ideal I of R $\Longleftrightarrow R_p$ is Gorenstein for every minimal prime p (e.g., if R is a domain).

If m = 1 we can obtain an isomorphism $\Omega(R) \cong I$ as <u>graded</u> modules, up to a shift in grading. This is in general false for m > 1. Take $R = k[x,y,z]/(xy,xz,yz)$, deg x = (1,0,0), deg y = (0,1,0) and deg z = (0,0,1). Then R is a C-M ring, the localization at every minimal prime is a field, and $\Omega(R) \cong (x-y, x-z)$, but there is no way of realizing $\Omega(R)$ as a homogeneous ideal. However, if m > 1 and R is a domain one can realize $\Omega(R) \cong I$ as graded modules up to a shift in grading.

12.9 THEOREM. If $\Omega(R) \cong I$ then R/I is Gorenstein and either I = R or dim R/I = dim R - 1.

12.10 THEOREM. Let $\theta_1,\ldots,\theta_d$ be an hsop for M and $S = k[\theta_1,\ldots,\theta_d]$. Then $\Omega(M) \cong \mathrm{Hom}_S(M,S)$.

The isomorphism here is as R-modules. There is a standard way of making $\mathrm{Hom}_S(M,S)$ into an R-module: if $x \in R$, $\phi \in \mathrm{Hom}_S(M,S)$ and $u \in M$, define $(x\phi)(u) = \phi(xu)$.

§13. <u>A final look at linear equations</u>

We shall now return for the last time to the rings R_Φ of linear diophantine equations. Recall that Φ is an $r \times n$ $\mathbb{Z}$-matrix of maximal rank, $E_\Phi = \{\beta \in \mathbb{N}^n | \Phi\beta = 0\}$ and $R_\Phi = kE_\Phi$, the monoid algebra of E_Φ over k. The following discussion could be extended to the modules $M_{\Phi,\alpha}$, but for simplicity we consider only R_Φ which is always Cohen-Macaulay. Assume there exists $\beta \in E_\Phi$ such that $\beta > 0$. Recall that $H^d(R_\Phi) \cong k\{x^\beta | \Phi\beta = 0,\ \beta < 0\}$ and that in general $\hat{\Omega}(M) = H^d(M)^\vee$.

13.1 COROLLARY. $\Omega(R_\Phi) \cong k\{x^\beta | \beta \in E_\Phi,\ \beta > 0\}$.

Thus, $\Omega(R_\Phi)$ is isomorphic to an ideal in R_Φ. Since R_Φ is a domain we know $\Omega(R_\Phi)$ can in fact be realized as a graded ideal, and the above corollary identifies this ideal.

13.2 COROLLARY. R_Φ is Gorenstein $\Leftrightarrow$ $\exists$ unique minimal $\beta > 0$ in E_Φ (i.e., if $\gamma > 0$, $\gamma \in E_\Phi$, then $\gamma - \beta \geq 0$).

13.3 COROLLARY. R_Φ is Gorenstein if $(1,1,\ldots,1) \in E_\Phi$.

The last result has a nice equivalent formulation in terms of invariant theory: if $T \subseteq SL_n(k)$ is a torus acting on $R = k[x_1,\ldots,x_n]$, then R^T is Gorenstein. In this connection we would like to mention the following conjecture of Hochster, Stanley and others: If $G \subseteq SL_n(k)$ is linearly reductive, then R^G is Gorenstein. This is known to be true for finite groups (Watanabe [Wat]), tori (just shown) and semisimple groups (Hochster and Roberts [H-R]). Also, R^G is known to be Cohen-Macaulay for any linearly reductive $G \subseteq GL_n(k)$ (Hochster and Roberts [H-R]).

Finally, consider again the algebra of magic squares. Let E_Φ be the set of $n \times n$ $\mathbb{N}$-matrices having equal line sums.

$$\begin{bmatrix} 1 & \cdots & 1 \\ \vdots & & \vdots \\ 1 & \cdots & 1 \end{bmatrix} \in E_\Phi, \text{ hence } R_\Phi \text{ is Gorenstein, hence } H_n(r) = (-1)^{n-1}H_n(-n-r).$$

Conversely, if the last equality is proved combinatorially, which can be done, then $F(R_\Phi,\lambda) = (-1)^d\lambda^\alpha F\left(R_\Phi,\frac{1}{\lambda}\right)$, which by Theorem 12.7 implies that R_Φ is Gorenstein since R_Φ is a domain. The same arguments go through also for symmetric magic squares.

CHAPTER II

THE FACE RING OF A SIMPLICIAL COMPLEX

§1. Elementary properties of the face ring

Let Δ be a finite simplicial complex on the vertex set $V = \{x_1,\ldots,x_n\}$. Recall that this means that Δ is a collection of subsets of V such that $F \subseteq G \in \Delta \Rightarrow F \in \Delta$ and $\{x_i\} \in \Delta$ for all $x_i \in V$. The elements of Δ are called faces. If $F \in \Delta$, then define $\dim F := |F| - 1$ and $\dim \Delta := \max_{F\in\Delta} (\dim F)$. Let $d = \dim \Delta + 1$ Given any field k we now define the face ring (or Stanley-Reisner ring) $k[\Delta]$ of the complex Δ.

1.1 DEFINITION. $k[\Delta] = k[x_1,\ldots,x_n]/I_\Delta$, where

$$I_\Delta = \left(x_{i_1}x_{i_2}\cdots x_{i_r} \mid i_1 < i_2 < \ldots < i_r,\ \{x_{i_1},x_{i_2},\ldots,x_{i_r}\} \notin \Delta \right) .$$

1.2 EXAMPLE. Consider the following plane projection of a triangulation of the 2-sphere

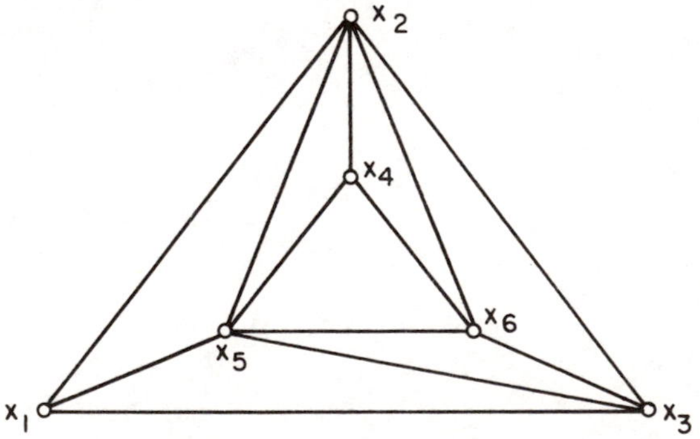

Here $I_\Delta = (x_1x_4, x_1x_6. x_3x_4, x_2x_5x_6, x_2x_3x_5)$.

1.3 **THEOREM.** $\dim k[\Delta] = 1 + \dim \Delta = d$.

Proof. $\dim k[\Delta]$ = maximal cardinality of an algebraically independent set of vertices $x_{i_1},\ldots,x_{i_j}$ = maximal cardinality of any face.□

Let f_i be the number of i-dimensional faces of Δ. Since $\emptyset \in \Delta$ and $\dim \emptyset = -1$, we get $f_{-1} = 1$. Also, $f_0 = |V|$. The d-tuple $f(\Delta) = (f_0,f_1,\ldots,f_{d-1})$ is called the f-vector of Δ. The theme for much of the following is to glean combinatorial information on f-vectors from algebraic information on face rings.

1.4 **THEOREM.** Define $\deg x_i = 1$. Then

$$H(k[\Delta],m) = \begin{cases} 1 & , \quad m = 0 \\ \sum_{i=0}^{d-1} f_i \binom{m-1}{i} & , \quad m > 0. \end{cases}$$

Equivalently,

$$F(k[\Delta],\lambda) = \sum_{i=-1}^{d-1} \frac{f_i\lambda^{i+1}}{(1-\lambda)^{i+1}} .$$

Note that the expression $\Sigma f_i\binom{m-1}{i}$ evaluated at $m = 0$ gives the Euler characteristic of Δ. Thus, the Hilbert function of $k[\Delta]$ lacks exceptional values if and only if $\chi(\Delta) = 1$. To prove the above it is easiest to work with a finer grading and then specialize. Define the fine grading of $k[\Delta]$ by $\deg x_i = (0,\ldots,0,\overset{i}{1},0,\ldots,0) \in \mathbb{Z}^n$, the i-th unit coordinate vector. Let $\operatorname{supp} x_1^{a_1}x_2^{a_2}\cdots x_n^{a_n} = \{x_i | a_i > 0\}$. Clearly, all monomials $u = x_1^{a_1}x_2^{a_2} \cdots x_n^{a_n}$ such that $\operatorname{supp} u \in \Delta$ form a k-basis for $k[\Delta]$. By counting such monomials u according to their support $F \in \Delta$ we arrive at the following expression for the Hilbert series of the fine grading:

$$F(k[\Delta],\lambda) = \sum_{F\in\Delta} \prod_{x_i\in F} \frac{\lambda_i}{1-\lambda_i},$$

Now replace all λ_i by λ to obtain Theorem 1.4.

§2. f-vectors and h-vectors of complexes and multicomplexes

What can be said in general about f-vectors of simplicial complexes? There is the following characterization given independently by Kruskal and Katona in response to a conjecture by Schützenberger (see [G-K] for references). Given two integers $\ell, i > 0$ write

$$\ell = \binom{n_i}{i} + \binom{n_{i-1}}{i-1} + \cdots + \binom{n_j}{j}, \quad n_i > n_{i-1} > \cdots > n_j \geq j \geq 1 .$$

A unique such expansion exists. Define

$$\ell^{(i)} = \binom{n_i}{i+1} + \binom{n_{i-1}}{i} + \cdots + \binom{n_j}{j+1} .$$

2.1 THEOREM (Kruskal, Katona). A vector $(f_0, f_1, \ldots, f_{d-1}) \in \mathbb{Z}^d$ is the f-vector of some (d-1)-dimensional simplicial complex Δ

$$\Longleftrightarrow 0 < f_{i+1} \leq f_i^{(i+1)}, \quad 0 \leq i \leq d-2 .$$

For instance, for i = 0: $f_1 \leq f_0^{(1)} \Longleftrightarrow f_1 \leq \binom{f_0}{2}$. Theorem 2.1 is proved using the following construction. List all i-element subsets $a_1 < a_2 < \cdots < a_i$ of $\mathbb{N}$ in reverse lexicographic order. For instance, for i = 3 the list starts 012, 013, 023, 123, 014, 024, 124, 034, 134, 234, 015, 025, ... Given $f = (f_0, f_1, \ldots, f_{d-1})$, $f_i > 0$, let $\Delta_f = \{\emptyset\} \cup \bigcup_{i=0}^{d-1}$ {first f_i (i+1)-element sets in above order}. One then verifies that the following are equivalent:

(i) f is the f-vector of a simplicial complex Δ,

(ii) Δ_f is a simplicial complex ,

(iii) $f_{i+1} \le f_i^{(i+1)}$, $i = 0,1,\dots,d-2$.

The difficult implication is (i) ⇒ (ii). For a nice proof of Theorem 2.1, see Greene-Kleitman [G-K, Sect. 8].

Along with simplicial complexes we shall need the more general notion of multicomplexes. A <u>multicomplex</u> Γ on $V = \{x_1,\dots,x_n\}$ is a set of monomials $x_1^{a_1}\cdots x_n^{a_n}$ such that $u \in \Gamma$, $v|u$ implies $v \in \Gamma$. So a simplicial complex corresponds to the case of squarefree monomials. Multicomplexes are sometimes called "semisimplicial complexes" by topologists. For a multicomplex Γ, let $h_i := \#\{u \in \Gamma | \deg u = i\}$, and define the <u>h-vector</u> $h(\Gamma) = (h_0,h_1,\dots)$. An h-vector may be infinite, and if $\Gamma \ne \emptyset$ then $h_0 = 1$. If $h_i = 0$ for $i > d$ we also write $h(\Gamma) = (h_0,\dots,h_d)$. A sequence $(h_0,h_1,\dots)$ which is the h-vector of some non-void multicomplex Γ will be called an <u>M-vector</u>.

Recall the definition of $\ell^{(i)}$, and define in analogy with the earlier notation

$$\ell^{<i>} = \binom{n_i+1}{i+1} + \binom{n_{i-1}+1}{i} + \cdots + \binom{n_j+1}{j+1} , \quad 0^{<i>} = 0 .$$

2.2 THEOREM (essentially Macaulay [Mac]). $(h_0,h_1,\dots)$ is an M-vector ⟺ $h_0 = 1$ and $0 \le h_{i+1} \le h_i^{<i>}$, $i \ge 1$.

Just as in the case of simplicial complexes, list all monomials of degree i in reverse lexicographic order. E.g., for i = 3:

$$x_1^3,\ x_1^2x_2,\ x_1x_2^2,\ x_2^3,\ x_1^2x_3,\ x_1x_2x_3,\ x_2^2x_3,\ x_1x_3^2,\ \dots$$

Given $h = (h_0,h_1,\dots)$ with $h_0 = 1$, let $\Gamma_h = \bigcup_{i\ge 0}$ {first h_i monomials of degree i in above order}. To prove Theorem 2.2, one then verifies that the following are equivalent:

(i) h is an M-vector

(ii) Γ_h is a multicomplex,

(iii) $0 \le h_{i+1} \le h_i^{<i>}$, $i \ge 1$.

Again, the difficult implication is (i) ⇒ (ii). Concerning the proof,

Macaulay [Mac, p.537] states: "The proof of the theorem...is given only to place it on record. It is too long and complicated to provide any but the most tedious reading." Macaulay's assessment is certainly accurate. Simpler proofs were given by Sperner, Whipple, and finally Clements and Lindström. Clements and Lindström [C-L] in fact prove a "generalized Macaulay theorem" which contains both the Kruskal-Katona and the Macaulay theorems as special cases.

The preceding enumerative considerations are closely related to the topic of Hilbert functions of graded algebras.

2.3 THEOREM (Macaulay [Mac]). Let R be an $\mathbb{N}$-graded k-algebra generated by $x_1,\ldots,x_n \in H(R_+)$. Then R has a k-basis which is a multicomplex on $\{x_1,\ldots,x_n\}$.

To prove this result one puts the monomials in $x_1,\ldots,x_n$ into reverse lexicographic order, selects a k-basis for R by taking the lexicographically earliest linearly independent subsequence of monomials, and then one shows that this basis forms a multicomplex. For details, see [St_6 Thm. 2.1].

2.4 COROLLARY. Fix a field k. Let $H:\mathbb{N}\to\mathbb{Z}$. Then the following are equivalent:

(i) $(H(0),H(1),\ldots)$ is an M-vector,

(ii) there exists a graded algebra $R = R_0 \oplus R_1 \oplus \cdots$, generated by R_1, such that $H(R,i) = H(i)$.

Proof. (ii) $\Rightarrow$ (i) is immediate from the previous theorem.

(i) $\Rightarrow$ (ii): Let Γ be a multicomplex on $\{x_1,\ldots,x_n\}$ such that $h(\Gamma) = (H(0),H(1),\ldots)$, and let $R = k[x_1,\ldots,x_n]/(x_1^{a_1}\cdots x_n^{a_n} \notin \Gamma)$. □

Now we define the h-vector of a graded algebra. Let $R = R_0 \oplus R_1 \oplus \cdots$ be generated by R_1, dim R = d. Thus,

$$F(R,\lambda) = \frac{h_0 + h_1\lambda + \cdots}{(1-\lambda)^d}\ , \quad h_0 + h_1\lambda + \cdots \in \mathbb{Z}[\lambda]\ .$$

Call the finite vector $h(R) = (h_0,h_1,\ldots)$ the <u>h-vector</u> of R.

2.5 COROLLARY. Fix $d \geq 0$. Let $(h_0,\ldots,h_\ell) \in \mathbb{Z}^{\ell+1}$. The following are equivalent:

(i) $(h_0,\ldots,h_\ell)$ is an M-vector

(ii) there exists a d-dimensional Cohen-Macaulay graded algebra $R = R_0 \oplus R_1 \oplus \cdots$, generated by R_1, such that $h(R) = (h_0,\ldots,h_\ell)$.

Proof. (i) $\Rightarrow$ (ii): Let $S = S_0 \oplus \cdots \oplus S_\ell$ be generated by S_1, with $H(S,i) = h_i$ (by the previous theorem). Let $R = S[x_1,\ldots,x_d]$. Then R is C-M; for a regular sequence of length d we can take $x_1,\ldots,x_d$; and when we mod out by them we are left with S.

(ii) $\Rightarrow$ (i): Let $\theta_1,\ldots,\theta_d$ be an hsop for R with deg $\theta_i = 1$. (Such a choice is always possible if the ground field k is infinite. If k is finite one may need to pass to an infinite extension field, which does not affect the Hilbert function or the Cohen-Macaulay property.) Let $S = R/(\theta_1,\ldots,\theta_d)$. Then $F(R,\lambda) = (1-\lambda)^{-d}F(S,\lambda)$, so $h_i = H(S,i)$ and $(h_0,\ldots,h_\ell)$ is an M-vector. □

Having previously defined the h-vectors of multicomplexes and graded rings, we now define the h-vector of a simplicial complex Δ as follows:

$$h(\Delta) := h(k[\Delta]) .$$

One observes that $h(\Delta)$ is a finite vector, in fact, $h_i = 0$ for $i > d = \dim \Delta + 1$. This definition is equivalent to the following explicit expression for $h(\Delta) = (h_0,h_1,\ldots,h_d)$ in terms of the f-vector $(f_0,f_1,\ldots,f_{d-1})$ of Δ (letting $f_{-1} = 1$):

$$h_k = \sum_{i=0}^{k} (-1)^{k-i}\binom{d-i}{k-i} f_{i-1} \quad , \quad 0 \le k \le d .$$

For example, if Δ is the boundary of an octahedraon then $f(\Delta) = (6,12,8)$ and $h(\Delta) = (1,3,3,1)$. Three observations are immediate:

(1) $h_d = (-1)^{d-1}\,\tilde{\chi}(\Delta)$, where $\tilde{\chi}$ denotes reduced Euler characteristic, i.e. $\tilde{\chi}(\Delta) = \sum_{i\ge -1} (-1)^i f_i = \sum_{i\ge -1} (-1)^i \dim_k \tilde{H}_i(\Delta;k)$,

(2) $h_0 + h_1 + \cdots + h_d = f_{d-1}$, and

(3) knowing the h-vector of Δ is equivalent to knowing the

f-vector of Δ.

§3. Cohen-Macaulay complexes and the Upper Bound Conjecture

3.1 DEFINITION. Δ is a Cohen-Macaulay complex (more precisely: C-M over k) if the face ring $k[\Delta]$ is Cohen-Macaulay.

3.2 COROLLARY. If Δ is Cohen-Macaulay, then $h(\Delta)$ is an M-vector.

Consider the following simplicial complex Δ:

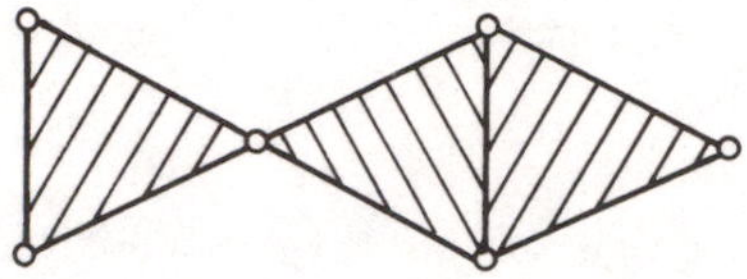

$f(\Delta) = (6,8,3)$, so $h(\Delta) = (1,3,-1,0)$. This is not an M-vector, so Δ is not Cohen-Macaulay.

A weak converse to Corollary 3.2 is also true, and we get the following result which can be called a "Kruskal-Katona theorem" for Cohen-Macaulay complexes. See [St_8, Thm. 6].

3.3 THEOREM. Let $h = (h_0, h_1, \ldots, h_d) \in \mathbb{Z}^{d+1}$. Then there exists a (d-1)-dimensional Cohen-Macaulay (or shellable [St_8]) complex Δ for which $h(\Delta) = h$ if and only if h is an M-vector. □

As a motivation for studying Cohen-Macaulay complexes we want to mention the Upper Bound Conjecture for spheres. This concerns the question: Suppose the geometric realization $|\Delta|$ is homeomorphic to the (d-1)-dimensional sphere S^{d-1}. Then given $f_0(\Delta) = n$, how large can f_i be? In order to formulate the conjectured answer we must review the notion of cyclic polytopes.

For $n > d$ let $C(n,d)$ be the convex hull of any n distinct points on the curve $\{(\tau, \tau^2, \ldots, \tau^d) \in \mathbb{R}^d \mid \tau \in \mathbb{R}\}$. The combinatorial type of the cyclic polytope $C(n,d)$ is independent of the n points chosen, and $\dim C(n,d) = d$. Cyclic polytopes have been investigated by Carathéodory, Gale, Motzkin, Klee, and others. Here are some known facts.

(a) $C(n,d)$ is simplicial (i.e., every proper face is a simplex), so the boundary $\partial C(n,d)$ defines an abstract simplicial complex $\Delta(n,d)$ such that $|\Delta(n,d)| \cong S^{d-1}$.

(b) $f_i(\Delta(n,d)) = \binom{n}{i+1}$ for $0 \le i < [\frac{d}{2}]$.

(c) $f_0, f_1, \ldots, f_{[\frac{d}{2}]-1}$ determine $f_{[\frac{d}{2}]}, f_{[\frac{d}{2}]+1}, \ldots, f_{d-1}$.

This is true for any Δ such that $|\Delta| \cong S^{d-1}$ because of the Dehn-Sommerville equations $h_i = h_{d-i}$, $0 \le i \le d$.

3.4 UPPER BOUND CONJECTURE (UBC) FOR SPHERES. If $|\Delta| \cong S^{d-1}$ and $f_0(\Delta) = n$, then $f_i(\Delta) \le f_i(\Delta(n,d))$, $i = 0,1,\ldots,d-1$.

McMullen [Mc_1] showed that Δ satisfies UBC if

$$h_i(\Delta) \le \binom{n-d+i-1}{i} \quad \text{for} \quad 0 \le i < [\tfrac{d}{2}] .$$

The conjectured upper bound for $f_i(\Delta)$ is true for $i < [\frac{d}{2}]$ by (b) and plausible for $i \ge [\frac{d}{2}]$ because of (c). The UBC for simplicial convex polytopes was proposed by Motzkin in 1957 [Mo] and proved by McMullen [Mc_1] in 1970. However, there exist triangulations of spheres, first found by Grünbaum, which are not polytopal. The smallest example of a non-polytopal sphere has parameters $d = 4$ and $f_0 = 8$. Klee suggested to extend the UBC to all spheres, and the general result was established by Stanley [St_4] as follows.

3.5 COROLLARY. If $|\Delta| \approx S^{d-1}$ and Δ is Cohen-Macaulay, then UBC holds for Δ.

Proof. $h_1(\Delta) = n - d$ and $(h_0, \ldots, h_d)$ is an M-vector. Hence, $h_i \le$ total number of monomials of degree i in n - d variables = $\binom{n-d+i-1}{i}$. □

To complete the proof of the UBC, we need to answer the question, when is a complex Δ Cohen-Macaulay? A very useful answer in terms of simplicial homology was given by Reisner in his 1974 thesis [R]. More general results were later obtained by Hochster [Ho_2]. In particular, the class of Cohen-Macaulay complexes can be seen to include all triangulations of spheres, so the UBC for spheres follows. We will now look at some of these results in more detail.

§4. Homological properties of face rings

Let Δ be a simplicial complex, and for $F \in \Delta$ define the link lk $F = \{G \in \Delta | G \cup F \in \Delta, G \cap F = \emptyset\}$. Recall that the fine grading of the face ring $k[\Delta]$ is given by setting $\deg x_i = (0,\ldots,0,1,0,\ldots,0) \in \mathbb{Z}^n$, the "1" in the i-th position.

4.1 THEOREM (Hochster, unpublished). Under the fine grading

$$F(H^i(k[\Delta]),\lambda) = \sum_{F \in \Delta} \dim_k \tilde{H}_{i-|F|-1}(\text{lk } F; k) \prod_{x_i \in F} \frac{\lambda_i^{-1}}{1 - \lambda_i^{-1}} .$$

The main goal of this section is to prove this fundamental result.

4.2 COROLLARY (Reisner [R]). Δ is Cohen-Macaulay over $k \iff$ for all $F \in \Delta$ and all $i < \dim(\text{lk } F)$, $\tilde{H}_i(\text{lk } F; k) = 0$.

Proof. By Hochster's theorem and Theorem 6.3 of Chapter I, $k[\Delta]$ is C-M $\iff \tilde{H}_{i-|F|-1}(\text{lk } F; k) = 0$ for all $i < d$ and all $F \in \Delta$. Assume that Δ is pure, i.e. that all maximal faces of Δ have dimension $d - 1 = \dim \Delta$. Then $\dim(\text{lk } F) = d - |F| - 1$ for all $F \in \Delta$, so $i < d \iff i - |F| - 1 < \dim(\text{lk } F)$. It only remains to verify that Δ is pure if either of the two conditions hold. Starting from Hochster's condition, if $|F| < d$ then $\tilde{H}_{-1}(\text{lk } F) = \tilde{H}_{|F|-|F|-1}(\text{lk } F) = 0$, so lk $F \neq \emptyset$ and F must be contained in a larger face. Starting from Reisner's condition one observes that the same condition holds for all proper links, hence by induction these are pure. If $\dim \Delta \geq 1$ then $\tilde{H}_0(\Delta) = \tilde{H}_0(\text{lk } \emptyset) = 0$, so Δ is connected, and this together with the purity of links of vertices shows that Δ is pure. □

The following result, due to Munkres [Mu], shows that the Cohen-Macaulayness of Δ is a topological property. The homology referred to in the second condition is reduced singular and relative singular homology respectively.

4.3 PROPOSITION (Munkres). Let $X = |\Delta|$. The following are equivalent:

(i) for all $F \in \Delta$ and all $i < \dim(\mathrm{lk}\, F)$, $H_i(\mathrm{lk}\, F; k) = 0$,

(ii) for all $p \in X$ and all $i < \dim X = d-1$,

$\tilde{H}_i(X;k) = H_i(X, X-p; k) = 0$.

4.4 COROLLARY. If $|\Delta| \cong S^{d-1}$ then Δ is Cohen-Macaulay.

4.5 COROLLARY. The UBC for spheres holds.

We will now prove Hochster's theorem (Theorem 4.1) on the local cohomology of face rings. Recall the formulation. Let Δ be a finite simplicial complex on vertices $x_1, x_2, \ldots, x_n$, and let k be a field. Give the face ring $R = k[\Delta]$ the fine grading $\deg x_i = (0,\ldots,0,1,0,\ldots,0) \in \mathbb{Z}^n$. Let $H^i(R)$ be the i-th local cohomology module with the induced $\mathbb{Z}^n$-grading. Then

$$F(H^i(R),\lambda) = \sum_{F\in\Delta} (\dim_k \tilde{H}_{i-|F|-1}(\mathrm{lk}\, F)) \prod_{x_i\in F} \frac{\lambda_i^{-1}}{1-\lambda_i^{-1}} .$$

Proof. The general plan for this proof is similar to what we did for the modules $M_{\Phi,\alpha}$. The idea is due to Hochster (unpublished), and was in fact our inspiration for Theorem 7.3 of the previous chapter.

Let $K(\underline{x}^\infty, R)$ be the complex of Chapter I, Theorem 6.2. Thus

$$H^i(R) = H^i(K(\underline{x}^\infty,R)) = H^i\Big(0 \xrightarrow{\delta_0} R \xrightarrow{\delta_1} \coprod_i R_{x_i} \xrightarrow{\delta_2} \coprod_{i<j} R_{x_ix_j} \xrightarrow{\delta_3} \cdots \xrightarrow{\delta_n} R_{x_1x_2\cdots x_n} \longrightarrow 0\Big)$$

$$= \ker \delta_{i+1} / \mathrm{im}\, \delta_i .$$

Let $F \subseteq \{x_1, x_2, \ldots, x_n\}$. Write R_F for R localized with respect to $\prod_{x_i\in F} x_i$. By inspection then

$$R_F = \begin{cases} k[\{x_i, x_i^{-1} | x_i \in F\} \cup \{x_j | x_j \in \text{lk } F\}] \ , & \text{if } F \in \Delta \\ 0 \ , & \text{if } F \notin \Delta \ . \end{cases}$$

The latter case is clear since if we invert a set whose product is zero then everything disappears. For the former case, notice that if $x_j \notin \text{st } F := \{G \in \Delta | G \cup F \in \Delta\}$ then $x_j \prod_{x_i \in F} x_i = 0$ in R. If $F \in \Delta$ we can equivalently write $R_F = k[\{x_i | x_i \in \text{st } F\} \cup \{x_i^{-1} | x_i \in F\}]$.

We want to compute $K(\underline{x}^\infty, R)_\alpha$ for $\alpha = (\alpha_1, \alpha_2, \ldots, \alpha_n) \in \mathbb{Z}^n$. First, if $\text{supp } \alpha := \{x_i | \alpha_i \neq 0\} \notin \Delta$ then $K(\underline{x}^\infty, R)_\alpha$ is zero. So suppose that $\text{supp } \alpha \in \Delta$ and let $F = \{x_i | \alpha_i < 0\}$ and $G = \{x_i | \alpha_i > 0\}$, $|F| = j$. Let us look at the r-th term in $K(\underline{x}^\infty, R)_\alpha$:

$$\left(\coprod_{i_1 < \cdots < i_r} R_{x_{i_1} \cdots x_{i_r}} \right)_\alpha = \left(\coprod_{\substack{F' \in \Delta \\ |F'| = r}} R_{F'} \right)_\alpha \ .$$

This is a vector space over k with basis corresponding to all $F' \supseteq F$ such that $|F'| = r$ and $F' \cup G \in \Delta$, i.e., (deleting F) to all (r-j)-element faces of $\text{lk}_{\text{st } G} F$ (the link within st G of F). The maps in $K(\underline{x}^\infty, R)_\alpha$ are Koszul relations. If we fix an orientation of Δ these are coboundary maps except possibly for signs. Thus, by choosing correct signs for the basis elements of each $(R_F)_\alpha$ we may identify $K(\underline{x}^\infty, R)_\alpha$ with the augmented oriented simplicial cochain complex of $\text{lk}_{\text{st } G} F$ with dimension shifted by $j + 1$. (Note that in the case of $M_{\Phi, \alpha}$ we got a chain complex, but here a cochain complex.) So,

$$H^i(R)_\alpha \cong \tilde{H}^{i-j-1}(\text{lk}_{\text{st } G} F) \cong \tilde{H}_{i-j-1}(\text{lk}_{\text{st } G} F) \ ,$$

(since we are working over a field).

4.6 EXAMPLE. Let Δ be the 1-dimensional complex:

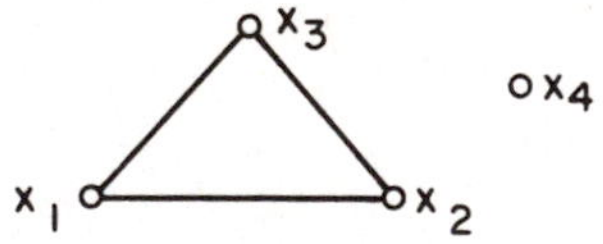

Then $K(\underline{x}^{\infty},R)$:

$$0 \to R \to R_{x_1} \oplus R_{x_2} \oplus R_{x_3} \oplus R_{x_4} \to R_{x_1x_2} \oplus R_{x_1x_3} \oplus R_{x_2x_3} \to 0 .$$

If $\alpha = (0,0,0,0)$, then $K(\underline{x}^{\infty},R)_\alpha$ is the augmented cochain complex for $\mathrm{lk}_{\mathrm{st}\,\emptyset}\,\emptyset = \Delta$ and

$$H^i(R)_\alpha = \tilde{H}_{i-1}(\Delta) .$$

If $\alpha = (-2,3,0,0)$, then $0 \to 0 \to (R_{x_1})_\alpha \to (R_{x_1x_2})_\alpha \to 0$ is the augmented cochain complex for $\mathrm{lk}_{\mathrm{st}\,x_2}\,x_1 = x_2$, and

$$H^i(R)_\alpha = \tilde{H}_{i-2}\,(\text{point}) .$$

Returning to the general proof, it now remains to sum over all $\alpha \in \mathbb{Z}^n$. First observe the following simplification. If $G \neq \emptyset$, then $\mathrm{lk}_{\mathrm{st}\,G}\,F$ is a cone over G and therefore acyclic. We may therefore assume that $G = \emptyset$, or equivalently $\alpha \leq 0$, so $\mathrm{lk}_{\mathrm{st}\,G}\,F = \mathrm{lk}_\Delta F = \mathrm{lk}\,F$ and $H^i(R)_\alpha \cong \tilde{H}_{i-|F|=1}\,(\mathrm{lk}\,F)$. Summing over all α we now get:

$$F(H^i(R),\lambda) = \sum_{F\in\Delta} \sum_{\substack{\alpha\leq 0 \\ \mathrm{supp}\,\alpha=F}} (\dim_k \tilde{H}_{i-|F|-1}\,(\mathrm{lk}\,F))\lambda^\alpha$$

$$= \sum_{F\in\Delta} \dim_k \tilde{H}_{i-|F|-1}\,(\mathrm{lk}\,F) \prod_{x_i\in F} \frac{\lambda_i^{-1}}{1-\lambda_i^{-1}} ,$$

and the proof is complete. □

Let Δ, $V = \{x_1,x_2,\ldots x_n\}$, $A = k[x_1,x_2,\ldots,x_n]$ and $k[\Delta] = A/I_\Delta$ be as before. Consider the $\mathbb{Z}^n$-graded finite-dimensional vector space $T_i = \mathrm{Tor}_i^A(k[\Delta],k)$. We would like to determine its Hilbert series. The answer, given by Hochster, implies a nice formula for the Betti numbers $\beta_i^A(k[\Delta]) = \dim_k \mathrm{Tor}_i^A(k[\Delta],k)$.

4.7 EXAMPLE. The complex Δ : 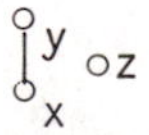

has a minimal free resolution

$$0 \longrightarrow A \xrightarrow{[y\ -x]} A^2 \xrightarrow{\begin{bmatrix} xz \\ yz \end{bmatrix}} A \longrightarrow k[\Delta] \longrightarrow 0$$

with degree-preserving maps and generators of degree (1,1,1); (1,0,1) and (0,1,1); and (0,0,0), respectively. Consequently,

$$F(T_0,\lambda) = 1 ,$$

$$F(T_1,\lambda) = \lambda_1\lambda_3 + \lambda_2\lambda_3 ,$$

$$F(T_2,\lambda) = \lambda_1\lambda_2\lambda_3 .$$

4.8 THEOREM (Hochster [Ho_2]).

$$F(T_i,\lambda) = \sum_{W \subseteq V} (\dim_k \tilde{H}_{|W|-i-1} (\Delta_W)) \prod_{x_i \in W} \lambda_i ,$$

where $\Delta_W = \{F \in \Delta | F \subseteq W\}$.

The proof is similar to but rather more complicated than the preceding one. Let us merely check that the formula works for the given example:

$$F(T_0,\lambda) = \dim_k \tilde{H}_{-1}(\emptyset) ,$$

$$F(T_1,\lambda) = \left(\dim_k \tilde{H}_0 \left(\begin{smallmatrix} x & z \\ \circ & \circ \end{smallmatrix}\right)\right) \lambda_1\lambda_3 + \left(\dim_k \tilde{H}_0 \left(\begin{smallmatrix} y & z \\ \circ & \circ \end{smallmatrix}\right)\right) \lambda_2\lambda_3 ,$$

$$F(T_2,\lambda) = \left(\dim_k \tilde{H}_0 \left(\begin{smallmatrix} \circ y \\ | \\ \circ x \end{smallmatrix} \quad \circ z\right)\right) \lambda_1\lambda_2\lambda_3 .$$

4.9 COROLLARY. $\beta_i^A(k[\Delta]) = \sum_{W \subseteq V} \dim_k \tilde{H}_{|W|-i-1} (\Delta_W)$

§5. Gorenstein face rings.

Let us now turn to this question: which CM complexes Δ are Gorenstein, i.e., for which Δ is the face ring $k[\Delta]$ a Gorenstein ring? Define for Δ on $V = \{x_1, x_2, \ldots, x_n\}$, core $V := \{x \in V \mid st\{x\} \neq V\}$, where $st\{x\} = \{F \in \Delta \mid F \cup \{x\} \in \Delta\}$, and

core $\Delta := \Delta_{\text{core } V}$.

Thus, $\Delta_{V\text{-core } V}$ is a simplex, and $\Delta = \Delta_{V\text{-core } V} *$ core Δ (where $\Delta' * \Delta'' = \{F \cup G \mid F \in \Delta', G \in \Delta''\}$ is the simplicial join). On the ring-theoretic level:

$$k[\Delta] = k[\text{core } \Delta][x \mid x \in V - \text{core } V] .$$

5.1 THEOREM. Fix a field k (or $\mathbb{Z}$). Let Δ be a simplicial complex and $\Gamma :=$ core Δ. Then the following are equivalent:

(a) Δ is Gorenstein (over k) ,

(b) for all $F \in \Gamma$, $\tilde{H}_i(lk_\Gamma F;k) = \begin{cases} k , & i = \dim(lk_\Gamma F) \\ 0 , & i < \dim(lk_\Gamma F) , \end{cases}$

(c) for $X = |\Gamma|$ and all $p \in X$,

$$\tilde{H}_i(X,k) \cong H_i(X,X-p;k) = \begin{cases} k , & i = \dim X \\ 0 , & i < \dim X , \end{cases}$$

(d) Δ is CM over k, and Γ is an orientable pseudomanifold over k (or even over $\mathbb{Z}$, regardless of k),

(e) either (i) $\Delta = \emptyset$, ○, or ○ ○, or
(ii) Δ is CM over k of dimension $d - 1 \geq 1$, and the link of every (d-3)-face is either a circle or ○—○ or ○—○—○, and $\tilde{\chi}(\Gamma) = (-1)^{\dim \Gamma}$ (the last condition is superfluous over $\mathbb{Z}$ or if char k = 2).

This characterization shows for instance that the 2-and 3-point lines, ○—○ and ○—○—○, are Gorenstein while the 4-point line ○—○—○—○ is not. Condition (e) was given by Hochster [Ho_2] (for $k = \mathbb{Z}$) and (b) - (c) by Stanley [St_8]. Condition (d) stems from a remark by Björner.

<u>Proof</u>. Let as before $T_i = \text{Tor}_i^A(k[\Delta],k)$, $\dim \Delta = d - 1$ and $n = |V|$. Then, $k[\Delta]$ is Gorenstein if and only if $F(T_{n-d+1},\lambda) = 0$

and $F(T_{n-d},\lambda) = \prod_{x_i \in W} \lambda_i$ for some $W \subseteq V$. One shows that $W = \text{core } V$ and then uses Hochster's formula for $F(T_{n-d},\lambda)$ to get (a) $\Leftrightarrow$ (b). The rest is an exercise in combinatorial topology. Let us merely make some remarks concerning condition (d). Suppose that Δ is CM over k and $\dim \Gamma = e - 1$. Then Γ is a pseudomanifold if and only if every (e-2)-dimensional face of Γ lies in exactly two (e-1)-dimensional faces. Furthermore, Γ is orientable over $k \Leftrightarrow \tilde{H}_{e-1}(\Gamma;k) \cong k \Leftrightarrow \tilde{\chi}(\Gamma) = (-1)^{e-1}$. In particular, Γ is orientable over $k \Leftrightarrow$ orientable over $\mathbb{Z}$. □

5.2 COROLLARY. If $|\Delta| \approx S^{d-1}$ then Δ is Gorenstein.

Let R be an $\mathbb{N}^m$-graded finitely-generated Gorenstein algebra. Preserve the usual notation and conventions, in particular $\text{hd}_A R = n-d$. Let $T_i = \text{Tor}_i^A(R,k)$ and $F(T_{n-d},\lambda) = \lambda^\alpha$. Then,

$$F(T_i,\lambda) = \lambda^\alpha F\left(T_{n-d-i}, \frac{1}{\lambda}\right),$$

where $\frac{1}{\lambda} = \left(\frac{1}{\lambda_1}, \frac{1}{\lambda_2}, \ldots, \frac{1}{\lambda_m}\right)$. Note that with $\lambda = 1$ this gives the relation $\beta^A_{n-d-i}(R) = \beta^A_i(R)$, which was earlier pointed out.

Now let Δ be a Gorenstein complex such that $\Delta = \text{core } \Delta$, and let us apply the preceding formula to $k[\Delta]$. In view of Hochster's formula we get

$$\sum_{W\subseteq V} \left(\dim_k \tilde{H}_{|W|-i-1}(\Delta_W)\right) \prod_{x_i \in W} \lambda_i$$

$$= \lambda_1\lambda_2\cdots\lambda_n \sum_{W\subseteq V} \left(\dim_k \tilde{H}_{|W|-n+d+i-1}(\Delta_W)\right) \prod_{x_i \in W} \lambda_i^{-1}$$

$$= \sum_{W\subseteq V} (\cdots) \prod_{x_i \notin W} \lambda_i .$$

Hence,

$$\tilde{H}_{|W|-i-1}(\Delta_W) \cong \tilde{H}_{|V-W|-n+d+i-1}(\Delta_{V-W}),$$

i.e., $\tilde{H}_j(\Delta_W) \cong \tilde{H}_{d-j-2}(\Delta_{V-W})$ for all $W \subseteq V$ and all j. This is the <u>Alexander duality theorem</u>.

§6. Gorenstein Hilbert functions

What can one say about Gorenstein Hilbert functions and about the h-vectors of Gorenstein complexes?

6.1 DEFINITION. A sequence $(h_0,h_1,\ldots,h_s)$, $h_s \neq 0$, is called Gorenstein if there exists a Gorenstein algebra $R = R_0 \oplus R_1 \oplus \cdots$ generated by R_1 such that $h(R) = (h_0,h_1,\ldots,h_s)$. (Recall that this means $F(R,\lambda) = (1-\lambda)^{-d}(h_0 + h_1\lambda + \cdots + h_s\lambda^s)$.)

PROBLEM 1. Characterize Gorenstein sequences.

Let Δ be a Gorenstein complex with dim (core Δ) = e - 1. Then the h-vector $h(\text{core}\,\Delta) = (h_0,h_1,\ldots,h_e)$ is a Gorenstein sequence and $h(\Delta) = (h_0,h_1,\ldots,h_e,0,\ldots,0)$. In contrast to the Cohen-Macaulay case (Theorem 3.3), not every Gorenstein sequence arises in this way (see fact (a) below).

PROBLEM 2. Characterize the h-vectors $h(\Delta)$ of Gorenstein complexes.

It has earlier been shown that every Gorenstein sequence $(h_0,h_1,\ldots,h_s)$ is symmetric: $h_i = h_{s-i}$. The equations $h_i = h_{s-i}$, $i = 0,1,\ldots,[\frac{s}{2}]$, for the h-vector $h(\Delta) = (h_0,h_1,\ldots,h_s)$ of a complex Δ are known as the Dehn-Sommerville equations.

6.2 CONJECTURE. $(h_0,h_1,\ldots,h_s)$ is the h-vector of some Gorenstein complex Δ = core Δ if and only if $h_i = h_{s-i}$ for all i and $(h_0, h_1 - h_0, h_2 - h_1,\ldots,h_{[s/2]}-h_{[s/2]-1})$ is an M-vector.

The sufficiency of the conjectured condition follows from a result of Billera and Lee [B-L]. The necessity is known if Δ is the boundary of a simplicial convex polytope (Stanley [St_{10}]) but is otherwise open.

Let us review some of the known facts concerning the two above-mentioned problems.

(a) The sequence (1,13,12,13,1) is Gorenstein (Stanley [St_6, p.70]) but not $h(\Delta)$ for some Gorenstein complex Δ (this follows from a result of Klee, cf. [St_8, p. 58]).

(b) Gorenstein sequences and Gorenstein $h(\Delta)$ agree for $h_1 \leq 3$ (follows from a result of Buchsbaum and Eisenbud, cf. [St_6, Thm. 4.2]).

(c) (1,4,3,4,1) and (1,5,4,5,1) are not Gorenstein sequences but are M-vectors (cf. [St_6, p. 71]).

Proof of (c). Let $R = R_0 \oplus R_1 \oplus \cdots \oplus R_s$ be a 0-dimensional graded Gorenstein algebra generated by R_1, with $R_s \neq 0$. Thus $R_s =$ soc R. If $x \in R_i$, then the principal ideal xR is isomorphic (as an R-module) to R/I for some ideal I. Since soc $xR = R_s$, there follows the well-known result that R/I is a Gorenstein ring. Hence if $a_j = \dim_k xR_j$, then $(a_0, a_1, \ldots, a_{s-i})$ is a Gorenstein sequence.

Now assume s = 4 and h(R) = (1,4,3,4,1). Let $x \in R_1$ and $xR \cong R/I$. Since R/I is Gorenstein, we have in the above notation $(a_0,a_1,a_2,a_3) = (1,a,a,1)$ for some $a \geq 1$, so that h(R/xR) = (1,3,3-a,4-a). But for no $a \geq 1$ is (1,3,3-a,4-a) an M-vector, so R does not exist.

Now assume h(R) = (1,5,4,5,1), and again let $x \in R_1$. Reasoning as above, we have h(R/xR) = (1,4,4-a,5-a). In order for this to be an M-vector, we must have a = 1. This means in particular that for every $x \in R_1$, $\dim_k xR_1 = 1$. Pick $x,y,z,w \in R_1$ such that xy and zw are linearly independent. (This is possible since $\dim_k R_2 > 1$.) Since $\dim_k xR_1 = 1$, we have $xz = \beta xy$ for some $\alpha \in k$. Similarly $xz = \beta zw$. Hence xz = 0. Similarly wy = 0. But since $\dim_k (x+w)R_1 = 1$ we have that (x+w)y and (x+w)z are linearly dependent. But (x+w)y = xy and (x+w)z = zw, a contradiction. □

(d) Let $f(n) = \min \{N \mid (1,n,N,n,1) \text{ is a Gorenstein sequence}\}$. Thus, f(4) = 4, f(5) = 5, and $f(13) \leq 12$. We claim:

$$\lim_{n\to\infty} \frac{\log f(n)}{\log n} = \frac{2}{3} .$$

Sketch of proof. (I am grateful to Peter Kleinschmidt for pointing out a gap in the original unpublished proof of this result.) Let $S = S_{(m)} = k[x_1, \cdots, x_m]/(x_1, \cdots, x_m)^4$. Let $R = R_{(m)}$ be the trivial extension of S by its injective envelope (as described in [St_6, p. 70]). Then R is Gorenstein, generated by R_1, and $h(R) = (1, \binom{m+2}{3} + m, 2\binom{m+1}{2}, \binom{m+2}{3} + m, 1)$. Let x be an indeterminate of degree one. Choose $0 \neq t \in \operatorname{soc} R$, and define $R^{[1]} = R[x]/I$, where I is the ideal of $R[x]$ generated by xR_1 and $x^4 - t$. Then $R^{[1]}$ is Gorenstein, generated by $R_1^{[1]}$, and $h(R^{[1]}) = h(R) + (0,1,1,1,0)$. Thus defining $R^{[j]} = \left(R^{[j-1]}\right)^{[1]}$, we have for all $j \geq 1$ that $R^{[j]}$ is Gorenstein, generated by $R_1^{[j]}$, and $h\left(R^{[j]}\right) = h(R) + (0,j,j,j,0)$. For any n, let m be the greatest integer such that $\binom{m+2}{3} + m \leq n$, and set $j = n - \binom{m+2}{3} - m$. Then the rings $T = R_{(m)}^{[j]}$ are Gorenstein, generated by T_1, and satisfy $h(T) = (1,n,t_n,n,1)$, where

$$\lim_{n\to\infty} \frac{\log t_n}{\log n} = \frac{2}{3} \quad .$$

(In fact, $\limsup t_n n^{-2/3} = \frac{3}{2} \cdot 6^{2/3}$.) This shows $\limsup \frac{\log f(n)}{\log n} \leq \frac{2}{3}$. But it follows easily just from the fact that $(1,n,f(n),n,1)$ is an M-vector that $\liminf \frac{\log f(n)}{\log n} \geq \frac{2}{3}$. (In fact, $\liminf f(n)n^{-2/3} \geq \frac{1}{2} \cdot 6^{2/3}$.) Thus the proof follows.

It is open whether $f(n)n^{-2/3}$ converges to some limit c, which from the above argument must satisfy $\frac{1}{2} \cdot 6^{2/3} \leq c \leq \frac{3}{2} \cdot 6^{2/3}$. We conjecture $c = 6^{2/3}$. Facts (c) and (d) were first stated in [St_6, Ex. 4.3] (with the bound $c \leq 6^{2/3}$ due to an erroneous argument), but no proofs were provided. P. Kleinschmidt has subsequently shown that

$$\limsup f(n)n^{-2/3} \leq 6^{2/3} ,$$

so if c exists then indeed $c \leq 6^{2/3}$.

Using the theory of ancestor ideals [I], one can give a purely linear algebraic definition (or characterization) of Gorenstein sequences. Let M_i denote the set of all monomials of degree i in the variables $x_1,\ldots,x_n$. Fix $s \in \mathbb{N}$ and a nonzero function $\sigma\colon M_s \to k$. For $0 \leq j \leq s$ define a matrix $A^{(j)}$, whose rows are indexed by M_j and columns by M_{s-j}, by the rule $A_{uv}^{(j)} = \sigma(uv)$. Let $h_j = \operatorname{rank} A^{(j)}$. Then $(h_0,h_1,\ldots,h_s)$ is a Gorenstein sequence (over k) with $h_1 \leq n$, and all such Gorenstein sequences arise in this way.

§7. Canonical modules of face rings

The next topic will be canonical modules of Cohen-Macaulay face rings. Recall (Theorem 12.8 of Chapter I) that $\Omega(R)$ is isomorphic to an ideal $I \subseteq R \Leftrightarrow R$ is generically Gorenstein (i.e., R_p is Gorenstein for all minimal primes p). Now, a face ring $k[\Delta]$ is generically a field. This suggests the following general problem: Imbed $\Omega(k[\Delta])$ as an ideal I of $k[\Delta]$, and describe $k[\Delta]/I$.

Recall that if R is a graded Cohen-Macaulay algebra of dimension d then $F(\Omega(R),\lambda) = (-1)^d F(R,\frac{1}{\lambda})$ (up to a shift in the grading of $\Omega(R)$). The right-hand side can be explicitly computed for any $R = k[\Delta]$.

7.1 THEOREM. Let Δ be any (d-1)-dimensional simplicial complex and give $k[\Delta]$ the fine grading. Then

$$(-1)^d F(k[\Delta],\tfrac{1}{\lambda}) = \sum_{F\in\Delta} (-1)^{d-|F|-1} \tilde{\chi}(\operatorname{lk} F) \prod_{x_i\in F} \frac{\lambda_i}{1-\lambda_i} .$$

<u>Proof</u>. $F(k[\Delta],\lambda) = \sum_{F\in\Delta} \prod_{x_i\in F} \frac{\lambda_i}{1-\lambda_i}$, so

$$(-1)^d F(k[\Delta],\tfrac{1}{\lambda}) = (-1)^d \sum_{F\in\Delta} (-1)^{|F|} \prod_{x_i\in F} \frac{1}{1-\lambda_i} .$$

Let $\alpha = (\alpha_1,\ldots,\alpha_n) \in \mathbb{N}^n$ and $F = \{x_i \mid \alpha_i > 0\} \in \Delta$. The coefficient of λ^α is $(-1)^d \sum_{\substack{G\in\Delta\\ G\supseteq F}} (-1)^{|G|} = (-1)^d \sum_{G'\in \operatorname{lk} F} (-1)^{|F|+|G'|} = (-1)^{d+|F|-1} \tilde{\chi}(\operatorname{lk} F)$. □

7.2 COROLLARY. Let $|\Delta|$ be a manifold with boundary (possibly void). Then

$$(-1)^d F(k[\Delta],\tfrac{1}{\lambda}) = (-1)^{d-1}\tilde{\chi}(\Delta) + \sum_{\substack{F\in\Delta-\partial\Delta \\ F\neq\emptyset}} \prod_{x_i\in F} \frac{\lambda_i}{1-\lambda_i} .$$

If $|\Delta|$ is a Cohen-Macaulay manifold of dimension ≥ 1 with nonvoid boundary then $\tilde{\chi}(\Delta) = 0$. Hence, in this case

$$F(\Omega(k[\Delta]),\lambda) = \sum_{F\in\Delta-\partial\Delta} \prod_{x_i\in F} \frac{\lambda_i}{1-\lambda_i} = F(I,\lambda) ,$$

where I is the ideal of $k[\Delta]$ generated by all $F \in \Delta - \partial\Delta$. Thus it is natural to ask whether $\Omega(k[\Delta]) \cong I$.

7.3 THEOREM (Hochster, unpublished). Let $|\Delta|$ be a Cohen-Macaulay manifold with nonvoid boundary $|\partial\Delta|$. Then

$$I \cong \Omega(k[\Delta]) \iff \partial\Delta \quad \text{is Gorenstein.}$$

(Remark: $\partial\Delta$ is Gorenstein e.g. if $|\Delta|$ is orientable.)

Proof. ($\Leftarrow$) Let C be a cone over $\partial\Delta$ with vertex y. Let $\Gamma = \Delta \cup C$, where Δ and C are identified along $\partial\Delta = \partial C$. $|\Gamma|$ is a manifold, except possibly at y where lk $y = \partial\Delta$. By an elementary application of the Mayer-Vietoris exact sequence, $\tilde{H}_d(\Gamma) \cong \tilde{H}_{d-1}(\partial\Delta) \cong k$ and $H_i(\Gamma) = 0$ for $i \neq d$. One similarly checks the proper links to find that Γ is Gorenstein. Since $S = k[\Gamma]$ is Gorenstein and $R = k[\Delta] = k[\Gamma]/J$ is Cohen-Macaulay of the same dimension d it follows that $\Omega(k[\Delta]) \cong \mathrm{Hom}_S(R,S)$ (it is a well-known fact that in general if S is Gorenstein and $R = S/J$ then $\Omega(R) \cong \mathrm{Ext}_{\dim S-\dim R}(R,S)$). Thus, $\Omega(k[\Delta]) \cong \mathrm{Ann}_S J = I$.

($\Rightarrow$) Recall (Theorem 12.9 of Chapter I) that in general if R is Cohen-Macaulay and for an ideal $I \cong \Omega(R)$ then R/I is Gorenstein. In the present situation $k[\Delta]/I \cong k[\partial\Delta]$, so if $I \cong \Omega(k[\Delta])$ then $\partial\Delta$ is Gorenstein. □

What can one say about canonical modules of manifolds without boundary? If $|\Delta|$ is a Cohen-Macaulay orientable manifold without boundary, then $k[\Delta]$ is Gorenstein so $\Omega(k[\Delta]) \cong k[\Delta]$. If $|\Delta|$ is a Cohen-Macaulay nonorientable manifold without boundary, then $\tilde{\chi}(\Delta) = 0$ so by our earlier computation $F(\Omega(k[\Delta]),\lambda) = F(k[\Delta]_+,\lambda)$ where $k[\Delta]_+ = (x_1,x_2,\ldots,x_n)$. One might first suspect that $\Omega(k[\Delta]) \cong k[\Delta]_+$, but this would contradict Theorem 12.9 of Chapter I. The actual description of $\Omega(k[\Delta])$ can be obtained as follows. Orient st $x_i = \{F \in \Delta | F \cup x_i \in \Delta\}$ for each $x_i \in V$. Define a 1-Cech cocycle σ as follows. If $\{x_i,x_j\} \in \Delta$, let

$$\sigma\{x_i,x_j\} = \begin{cases} 1, & \text{if st } x_i \text{ and st } x_j \text{ have compatible orientations} \\ & \quad \text{(i.e., agree on intersection)}, \\ -1, & \text{if not.} \end{cases}$$

(Remark: σ is the obstruction cocycle for orientability; the image of σ in $H^1(\Delta)$ is $0 \Longleftrightarrow \Delta$ is orientable.) Paste together the ideals $x_ik[\Delta]$ as follows: identify $x_ix_jk[x] \subseteq x_ik[\Delta]$ with $\sigma\{x_i,x_j\}x_ix_jk[\Delta] \subseteq x_jk[\Delta]$. In other words, form $M = \coprod_i x_ik[\Delta]/(x_je_i - \sigma\{x_i,x_j\}x_ie_j)$ where e_i is the image of x_i in the i-th summand.

7.4 THEOREM (Hochster, unpublished). $M \cong \Omega(k[\Delta])$.

The proof is omitted.

The problem of giving a "nice" description of $\Omega(k[\Delta])$ for arbitrary Cohen-Macaulay Δ remains open. One additional case of interest was worked out by Baclawski. Define Δ to be doubly Cohen-Macaulay (2-CM) if Δ is Cohen-Macaulay, and for every vertex x of Δ the subcomplex $\Delta\backslash x = \{F \in \Delta : x \notin F\}$ is Cohen-Macaulay of the same dimension as Δ. Walker [Wal] has shown that double-Cohen-Macaulayness is a topological property, i.e., depends only on $|\Delta|$. For instance, spheres are 2-CM but cells are not. If Δ is CM, then Δ is 2-CM if and only if $(-1)^{d-1}\tilde{\chi}(\Delta) = \text{type } k[\Delta]$ (see [B_1]). For any Δ we may identify $F \in \Delta$ with $\Pi_{x \in F}\, x \in k[\Delta]$. In this way the augmented oriented chain complex of Δ (over k) can be imbedded (as a vector space) in $k[\Delta]$. In particular, if dim Δ = d-1 then the reduced homology group $\tilde{H}_{d-1}(\Delta;k)$ is imbedded in $k[\Delta]$, since it is a subspace of the (d-1)-chains.

7.5 THEOREM (Baclawski [B_2]). Suppose Δ is 2-CM of dimension d-1 and <u>balanced</u>, i.e., its vertex set V can be partitioned into disjoint subsets $V_1, V_2, \ldots, V_d$ such that each maximal face of Δ meets each V_i in one point. Then $\Omega(k[\Delta])$ is isomorphic to the ideal of $k[\Delta]$ generated by $\tilde{H}_{d-1}(\Delta;k)$. □

Theorem 7.5 is undoubtedly true without the assumption that Δ is balanced, but this remains to be proved.

If M is any Cohen-Macaulay module then $\Omega^2(M) := \Omega(\Omega(M))$ is isomorphic to M. The non-Cohen-Macaulay case was worked out by Hochster. Here we define $\Omega(M) = \mathrm{Ext}_A^{n-d}(M,A)$, where M is finitely-generated over the Gorenstein ring A, and where n = dim A, d = dim M. For face rings $k[\Delta]$ it has the following informal description. First "purify" Δ by removing all faces not contained in a face of maximum dimension d - 1. This yields a pure simplicial complex Δ'. Next choose a nonvoid face F of Δ' of smallest possible dimension $\leq$ d - 3 such that lk F is not connected, and "pull apart" Δ' at F by creating a copy F_i of F for each connected component K_i of lk F, so that lk $F_i = K_i$.

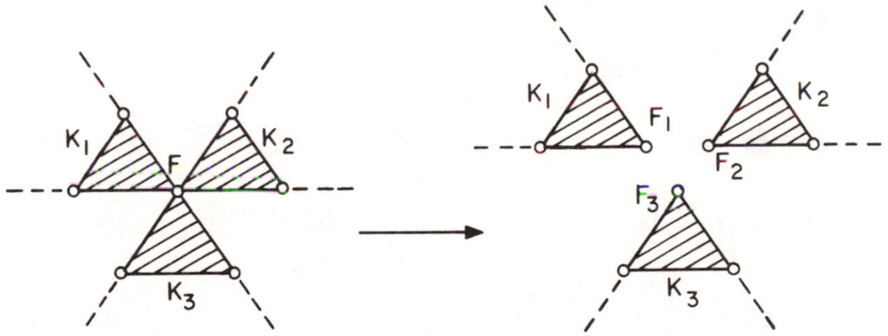

Find another nonvoid face F' of smallest possible dimension $\leq$ d-3 such that lk F' is not connected, and repeat the procedure. Continue until the link of every nonvoid face of dimension $\leq$ d-3 is connected. Let $\Delta_1, \Delta_2, \ldots, \Delta_j$ be the connected components of the resulting simplicial complex. Then

$$\Omega^2(k[\Delta]) \cong \coprod_{i=1}^{j} k[\Delta_i] ,$$

as a $k[x_1,\ldots,x_n]$-module or $k[\Delta]$-module. In particular, $\Omega^2(k[\Delta]) \cong k[\Delta]$ if and only if Δ is pure and the link of every face (including $\emptyset$) of dimension $\leq d-3$ is connected.

§8. Buchsbaum complexes

The final topic will be a brief glimpse of Buchsbaum complexes. Let R be an $\mathbb{N}$-graded finitely-generated algebra and let M be a d-dimensional $\mathbb{Z}$-graded finitely-generated R-module. M is said to be Buchsbaum if for every hsop $\theta_1,\ldots,\theta_d$ and all $1 \leq i \leq d$,

$$\{u \in M/(\theta_1 M + \cdots + \theta_{i-1}M) \mid u\theta_i = 0\} = \operatorname{soc} M/(\theta_1 M + \cdots + \theta_{i-1}M).$$

When applied to face rings $k[\Delta]$ this notion carries over to simplicial complexes Δ.

8.1 THEOREM (Schenzel [Sc]). Let Δ be a finite simplicial complex and k a field. Then the following are equivalent:

(i) Δ is Buchsbaum over k,

(ii) $k[\Delta]_p$ is Cohen-Macaulay for all primes $p \neq k[\Delta]_+$,

(iii) for all $F \in \Delta$, $F \neq \emptyset$ and $i < \dim(\operatorname{lk} F)$ imply $\tilde{H}_i(\operatorname{lk} F;k) = 0$,

(iv) for all $p \in X = |\Delta|$ and $i < \dim X$, $H_i(X,X-p;k) = 0$,

(v) $\dim_k H^i(k[\Delta]) < \infty$ if $0 \leq i < \dim k[\Delta] = d$ (in which case $H^i(k[\Delta]) \cong \tilde{H}_{i-1}(\Delta;k)$, $i < d$).

Remark: A characterization such as (v) is not known for general Buchsbaum rings.

8.2 THEOREM (Schenzel [Sc]). Let Δ be a Buchsbaum complex, let $\deg x_i = 1$ and let $\theta_1,\ldots,\theta_d \in k[\Delta]_1$ be an hsop. Then

$$(1-\lambda)^d F(k[\Delta],\lambda) = F(k[\Delta]/(\theta_1,\ldots,\theta_d),\lambda)$$

$$- \sum_{j=1}^{d} \binom{d}{j}\left(\sum_{i=0}^{j-1} (-1)^{j-i-1} \dim_k \tilde{H}_{i-1}(\Delta)\right)\lambda^j .$$

The last term can be interpreted as measuring the error when we leave the Cohen-Macaulay case. Set

$$h_0 + h_1\lambda + \cdots + h_d\lambda^d := (1-\lambda)^d F(k[\Delta],\lambda) \text{ , and}$$

$$g_0 + g_1\lambda + \cdots + g_d\lambda^d := F(k[\Delta]/(\theta_1,\ldots,\theta_d),\lambda) .$$

Then $g_1 = n - d$ and $g_j \le \binom{n-d+j-1}{j}$, the number of monomials of degree j in n-d variables. Hence,

$$h_j \le \binom{n-d+j-1}{j} - (-1)^j\binom{d}{j} \sum_{i=-1}^{j-2} (-1)^i \dim_k \tilde{H}_i(\Delta) . \qquad (5)$$

This inequality for the h-vector of a Buchsbaum complex Δ generalizes the inequality for a CM complex which yields the Upper Bound Conjecture. (It should be noted, however, that the Upper Bound Conjecture does not follow formally from (5); one also needs the Dehn-Sommerville equations.) It implies the following bounds on the f-vector $f(\Delta) = (f_0,\ldots,f_{d-1})$:

$$f_j \le \binom{n}{j} - \binom{d}{j} \sum_{i=-1}^{j-2} \binom{j-1}{i+1} \dim_k \tilde{H}_i(\Delta) .$$

If in addition to being Buchsbaum Δ is an orientable homology manifold (i.e., $\tilde{H}_{\dim(\mathrm{lk}\, F)}(\mathrm{lk}\, F) \cong k$ for all $F \in \Delta$), then $h_i = h_{d-i} + (-i)^d\binom{d}{i}((-1)^{d-1}-\tilde{\chi}(\Delta))$. Note that if d is even then $\tilde{\chi}(\Delta) = -1$ due to Poincaré duality, hence in this case $h_i = h_{d-i}$.

References

[A-D-G] H. Anand, V.C. Dumir, and H. Gupta, A combinatorial distribution problem, Duke Math. J. 33 (1966), 757-769.

[A-M] M.F. Atiyah and I.G. Macdonald, Introduction to Commutative Algebra, Addison-Wesley, Reading, Massachusetts, 1969.

[B_1] K. Baclawski, Cohen-Macaulay connectivity and geometric lattices, European J. Combinatorics 3 (1982), 293-305.

[B_2] K. Baclawski, Canonical modules of partially ordered sets, Report No. 8, Institut Mittag-Leffler, 1981.

[B-L] L.J. Billera and C.W. Lee, Sufficiency of McMullen's condition for f-vectors of simplicial polytopes, Bull. Amer. Math. Soc. 2 (1980), 181-185.

[C-L] G. Clements and B. Lindström, A generalization of a combinatorial theorem of Macaulay, J. Combinatorial Theory 7 (1969), 230-238.

[E_1] E. Ehrhart, Sur un problème de géometrie diophantienne linéaire, I, II, Crelle J. (= J. Reine Angew. Math.) 226 (1967), 1-29; 227 (1967), 25-49.

[E_2] E. Ehrhart, Démonstration de la loi de réciprocité du polyèdre rationnel, C.R. Acad. Sci. Paris 265A (1967), 91-94.

[F] J. Folkman, The homology groups of a lattice, J. Math. Mech. 15 (1966), 631-636.

[G-K] C. Greene and D.J. Kleitman, Proof techniques in the theory of finite sets, in Studies in Combinatorics (G.-C. Rota, ed.), Mathematical Association of America, 1978, pp. 22-79.

[Ha] R. Hartshorne, Local Cohomology, Lecture Notes in Math., no. 41, Springer, Berlin-Heidelberg-New York, 1967.

[H-K] J. Herzog and E. Kunz (eds.), Der kanonische Modul eines Cohen-Macaulay-Rings, Lecture Notes in Math., no. 238, Springer, Berlin-Heidelberg-New York, 1971.

[Ho_1] M. Hochster, Rings of invariants of tori, Cohen-Macaulay rings generated by monomials, and polytopes, Annals of Math. 96 (1972), 318-337.

[Ho_2] M. Hochster, Cohen-Macaulay rings, combinatorics, and simplicial complexes, in Ring Theory II (Proc. Second Oklahoma Conference) (B.R. McDonald and R. Morris, ed.), Dekker, New York, 1977, pp. 171-223.

[H-R] M. Hochster and J.L. Roberts, Rings of invariants of reductive groups acting on regular rings are Cohen-Macaulay, Advances in Math. 13 (1974), 115-175.

[I] A. Iarrobino, Vector spaces of forms I: Ancestor ideals, preprint.

[J-V] D.M. Jackson and G.H.J. van Rees, The enumeration of generalized double stochastic nonnegative integer square matrices, SIAM J. Comput. 4 (1975), 474-477.

[Mac] F.S. Macaulay, Some properties of enumeration in the theory of modular systems, Proc. London Math. Soc. 26 (1927), 531-555.

[Md_1] I.G. Macdonald, The volume of a lattice polyhedron, Proc. Cambridge Philos. Soc. 59 (1963), 719-726.

[Md_2] I.G. Macdonald, Polynomials associated with finite cell complexes, J. London Math. Soc. (2) 4 (1971), 181-192.

[MM] P.A. MacMahon, Combinatory Analysis, vols. 1-2, Cambridge, 1916; reprinted by Chelsea, New York, 1960.

[Mc_1] P. McMullen, The maximum numbers of faces of a convex polytope, Mathematika 17 (1970), 179-184.

[Mc_2] P. McMullen, Valuations and Euler-type relations for certain classes of convex polytopes, Proc. London Math. Soc. 35 (1977), 113-135.

[Mo] T.S. Motzkin, Comonotone curves and polyhedra, Abstract 111, Bull. Amer. Math. Soc. 63 (1957), 35.

[Mu] J. Munkres, Topological results in combinatorics, preprint.

[Pi] G. Pick, Geometrisches zur Zahlenlehre, Naturwissenschaft Zeitschrift Lotos, Prague, 1899.

[Po] T. Popoviciu, Asupra unei probleme de partitie a numerelor, Studie şi cercetari ştiintifice, Akad. R.P.R. Filiala Cluj 4 (1953), 7-58.

[R] G. Reisner, Cohen-Macaulay quotients of polynomial rings, Advances in Math. 21 (1976), 30-49.

[Sc] P. Schenzel, On the number of faces of simplicial complexes and the purity of Frobenius, Math. Z. 178 (1981), 125-142.

[St_1] R. Stanley, Linear homogeneous diophantine equations and magic labelings of graphs, Duke Math. J. 40 (1973), 607-632.

[St_2] R. Stanley, Combinatorial reciprocity theorems, Advances in Math. 14 (1974), 194-253.

[St_3] R. Stanley, Cohen-Macaulay rings and constructible polytopes, Bull. Amer. Math. Soc. 81 (1975), 133-135.

[St_4] R. Stanley, The Upper Bound Conjecture and Cohen-Macaulay rings, Studies in Applied Math. 54 (1975), 135-142.

[St_5] R. Stanley, Generating functions, in Studies in Combinatorics (G.-C. Rota, ed.), Mathematical Assoc. of America, 1978, pp. 100-141.

[St_6] R. Stanley, Hilbert functions of graded algebras, Advances in Math. 28 (1978), 57-83.

[St_7] R. Stanley, Magic labelings of graphs, symmetric magic squares, systems of parameters, and Cohen-Macaulay rings, Duke Math. J. 43 (1976), 511-531.

[St_8] R. Stanley, Cohen-Macaulay complexes, in Higher Combinatorics (M. Aigner, ed.), Reidel, Dordrecht and Boston, 1977, pp. 51-62.

[St_9] R. Stanley, Decompositions of rational convex polytopes, Annals of Discrete Math. 6 (1980), 333-342.

[St_{10}] R. Stanley, The number of faces of a simplicial convex polytope, Advances in Math. 35 (1980), 236-238.

[St_{11}] R. Stanley, Linear diophantine equations and local cohomology, Inv. math. 68 (1982), 175-193.

[Wal] J.W. Walker, Topology and combinatorics of ordered sets, thesis, M.I.T., 1981.

[Wat] K. Watanabe, Certain invariant subrings are Gorenstein. I, Osaka J. Math. 11 (1974), 1-8.

Progress in Mathematics
Edited by J. Coates and S. Helgason

Progress in Physics
Edited by A. Jaffe and D. Ruelle

- A collection of research-oriented monographs, reports, notes arising from lectures or seminars
- Quickly published concurrent with research
- Easily accessible through international distribution facilities
- Reasonably priced
- Reporting research developments combining original results with an expository treatment of the particular subject area
- A contribution to the international scientific community: for colleagues and for graduate students who are seeking current information and directions in their graduate and post-graduate work.

Manuscripts

Manuscripts should be no less than 100 and preferably no more than 500 pages in length.

They are reproduced by a photographic process and therefore must be typed with extreme care. Symbols not on the typewriter should be inserted by hand in indelible black ink. Corrections to the typescript should be made by pasting in the new text or painting out errors with white correction fluid.

The typescript is reduced slightly (75%) in size during reproduction; best results will not be obtained unless the text on any one page is kept within the overall limit of 6x9½ in (16x24 cm). On request, the publisher will supply special paper with the typing area outlined.

Manuscripts should be sent to the editors or directly to:
Birkhäuser Boston, Inc., P.O. Box 2007, Cambridge, Massachusetts 02139

PROGRESS IN MATHEMATICS
Already published

PM 1 Quadratic Forms in Infinite-Dimensional Vector Spaces
Herbert Gross
ISBN 3-7643-1111-8, 432 pages, paperback

PM 2 Singularités des systèmes différentiels de Gauss-Manin
Frédéric Pham
ISBN 3-7643-3002-3, 346 pages, paperback

PM 3 Vector Bundles on Complex Projective Spaces
C. Okonek, M. Schneider, H. Spindler
ISBN 3-7643-3000-7, 396 pages, paperback

PM 4 Complex Approximation, Proceedings, Quebec, Canada, July 3-8, 1978
Edited by Bernard Aupetit
ISBN 3-7643-3004-X, 128 pages, paperback

PM 5 The Radon Transform
Sigurdur Helgason
ISBN 3-7643-3006-6, 207 pages, hardcover

PM 6 The Weil Representation, Maslov Index and Theta Series
Gérard Lion, Michèle Vergne
ISBN 3-7643-3007-4, 348 pages, paperback

PM 7 Vector Bundles and Differential Equations
Proceedings, Nice, France, June 12-17, 1979
Edited by André Hirschowitz
ISBN 3-7643-3022-8, 256 pages, paperback

PM 8 Dynamical Systems, C.I.M.E. Lectures, Bressanone, Italy, June 1978
John Guckenheimer, Jürgen Moser, Sheldon E. Newhouse
ISBN 3-7643-3024-4, 305 pages, hardcover

PM 9 Linear Algebraic Groups
T. A. Springer
ISBN 3-7643-3029-5, 314 pages, hardcover

PM 10 Ergodic Theory and Dynamical Systems I
A. Katok
ISBN 3-7643-3036-8, 346 pages, hardcover

PM 11 18th Scandinavian Congress of Mathematicians, Aarhus, Denmark, 1980
Edited by Erik Balslev
ISBN 3-7643-3040-6, 526 pages, hardcover

PM 12 Séminaire de Théorie des Nombres, Paris 1979-80
Edited by Marie-José Bertin
ISBN 3-7643-3035-X, 404 pages, hardcover

PM 13 Topics in Harmonic Analysis on Homogeneous Spaces
Sigurdur Helgason
ISBN 3-7643-3051-1, 152 pages, hardcover

PM 14 Manifolds and Lie Groups, Papers in Honor of Yozô Matsushima
Edited by J. Hano, A. Marimoto, S. Murakami, K. Okamoto, and H. Ozeki
ISBN 3-7643-3053-8, 476 pages, hardcover

PM 15 Representations of Real Reductive Lie Groups
David A. Vogan, Jr.
ISBN 3-7643-3037-6, 776 pages, hardcover

PM 16 Rational Homotopy Theory and Differential Forms
Phillip A. Griffiths, John W. Morgan
ISBN 3-7643-3041-4, 258 pages, hardcover

PM 17 Triangular Products of Group Representations and their Applications
S. M. Vovsi
ISBN 3-7643-3062-7, 142 pages, hardcover

PM 18 Géométrie Analytique Rigide et Applications
Jean Fresnel, Marius van der Put
ISBN 3-7643-3069-4, 232 pages, hardcover

PM 19 Periods of Hilbert Modular Surfaces
Takayuki Oda
ISBN 3-7643-3084-8, 144 pages, hardcover

PM 20 Arithmetic on Modular Curves
Glenn Stevens
ISBN 3-7643-3088-0, 236 pagers, hardcover

PM 21 Ergodic Theory and Dynamical Systems II
A. Katok, editor
ISBN 3-7643-3096-1, 226 pages, hardcover

PM 22 Séminaire de Théorie des Nombres, Paris 1980-81
Marie-José Bertin, editor
ISBN 3-7643-3066-X, 374 pages, hardcover

PM 23 Adeles and Algebraic Groups
A. Weil
ISBN 3-7643-3092-9, 138 pages, hardcover

PM 24 Enumerative Geometry and Classical Algebraic Geometry
Patrick Le Barz, Yves Hervier, editors
ISBN 3-7643-3106-2, 260 pages, hardcover

PM 25 Exterior Differential Systems and the Calculus of Variations
Phillip A. Griffiths
ISBN 3-7643-3103-8, 349 pages, hardcover

PM 26 Number Theory Related to Fermat's Last Theorem
Neal Koblitz, editor
ISBN 3-7643-3104-6, 376 pages, hardcover

PM 27 Differential Geometric Control Theory
Roger W. Brockett, Richard S. Millman, Hector J. Sussmann, editors
ISBN 3-7643-3091-0, 349 pages, hardcover

PM 28 Tata Lectures on Theta I
David Mumford
ISBN 3-7643-3109-7, 254 pages, hardcover

PM 29 Birational Geometry of Degenerations
Robert Friedman and David R. Morrison, editors
ISBN 3-7643-3111-9, 410 pages, hardcover

PM 30 CR Submanifolds of Kaehlerian and Sasakian Manifolds
Kentaro Yano, Masahiro Kon
ISBN 3-7643-3119-4, 223 pages, hardcover

PM 31 Approximations Diophantiennes et Nombres Transcendants
D. Bertrand and M. Waldschmidt, editors
ISBN 3-7643-3120-8, 349 pages, hardcover

PM 32 Differential Geometry
Robert Brooks, Alfred Gray, Bruce L. Reinhart, editors
ISBN 3-7643-3134-8, 267 pages, hardcover

PM 33 Uniqueness and Non-Uniqueness in the Cauchy Problem
Claude Zuily
ISBN 3-7643-3121-6, 185 pages, hardcover

PM 34 Systems of Microdifferential Equations
Masaki Kashiwara
ISBN 0-8176-3138-0
ISBN 3-7643-3138-0, 182 pages, hardcover

PM 35 Arithmetic and Geometry Papers Dedicated to I. R. Shafarevich on the Occasion of His Sixtieth Birthday Volume I Arithmetic
Michael Artin, John Tate, editors
ISBN 3-7643-3132-1, 373 pages, hardcover

PM 36 Arithmetic and Geometry Papers Dedicated to I. R. Shafarevich on the Occasion of His Sixtieth Birthday Volume II Geometry
Michael Artin, John Tate, editors
ISBN 3-7643-3133-X, 495 pages, hardcover

PM 37 Mathématique et Physique
Louis Boutet de Monvel, Adrien Douady, Jean-Louis Verdier, editors
ISBN 0-8176-3154-2
ISBN 3-7643-3154-2, 454 pages, hardcover

PM 38 Séminaire de Théorie des Nombres, Paris 1981-82
Marie-José Bertin, editor
ISBN 0-8176-3155-0
ISBN 3-7643-3155-0, 359 pages, hardcover

PM 39 Classical Algebraic and Analytic Manifolds
Kenji Ueno, editor
ISBN 0-8176-3137-2
ISBN 3-7643-3137-2, 644 pages, hardcover

PM 40 Representation Theory of Reductive Groups
P. C. Trombi, editor
ISBN 0-8176-3135-6
ISBN 3-7643-3135-6, 308 pages, hardcover

PROGRESS IN PHYSICS

Already published

PPh 1 Iterated Maps on the Interval as Dynamical Systems
Pierre Collet and Jean-Pierre Eckmann
ISBN 3-7643-3026-O, 256 pages, hardcover

PPh 2 Vortices and Monopoles, Structure of Static Gauge Theories
Arthur Jaffe and Clifford Taubes
ISBN 3-7643-3025-2, 294 pages, hardcover

PPh 3 Mathematics and Physics
Yu. I. Manin
ISBN 3-7643-3027-9, 112 pages, hardcover

PPh 4 Lectures on Lepton Nucleon Scattering and Quantum Chromodynamics
W. B. Atwood, J. D. Bjorken, S. J. Brodsky, and R. Stroynowski
ISBN 3-7643-3079-1, 574 pages, hardcover

PPh 5 Gauge Theories: Fundamental Interactions and Rigorous Results
P. Dita, V. Georgescu, R. Purice, editors
ISBN 3-7643-3095-3, 406 pages, hardcover

PPh 6 Third Workshop on Grand Unification
University of North Carolina, Chapel Hill, April 15-17, 1982
P. H. Frampton, S. L. Glashow, and H. van Dam, editors
ISBN 3-7643-3105-4, 382 pages, hardcover